AF388990

Comparative Pathology and Immunohistochemistry of Veterinary Species

Comparative Pathology and Immunohistochemistry of Veterinary Species

Editors

Alejandro Suárez-Bonnet
Gustavo A. Ramírez

MDPI • Basel • Beijing • Wuhan • Barcelona • Belgrade • Manchester • Tokyo • Cluj • Tianjin

Editors
Alejandro Suárez-Bonnet
University of London
London, UK

Gustavo A. Ramírez
Universitat de Lleida
Lleida, Spain

Editorial Office
MDPI
St. Alban-Anlage 66
4052 Basel, Switzerland

This is a reprint of articles from the Special Issue published online in the open access journal *Animals* (ISSN 2076-2615) (available at: https://www.mdpi.com/journal/animals/special_issues/vet_pathoimmune).

For citation purposes, cite each article independently as indicated on the article page online and as indicated below:

LastName, A.A.; LastName, B.B.; LastName, C.C. Article Title. *Journal Name* **Year**, *Volume Number*, Page Range.

ISBN 978-3-0365-7994-8 (Hbk)
ISBN 978-3-0365-7995-5 (PDF)

Cover image courtesy of Gustavo A. Ramírez

Contents

About the Editors

Alejandro Suárez-Bonnet

Dr Alejandro Suárez-Bonnet is a veterinary surgeon with over 18 years of experience in pathology, including regular forensic examinations. He graduated as a Doctor in Veterinary Medicine (DVM) in 2005 from the University of Las Palmas de Gran Canaria (ULPGC), Spain, and received his PhD in animal health and comparative pathology after writing a thesis on cell markers in canine mammary tumours. He won the prize for the best PhD thesis in medical sciences in 2011. He worked as a pathologist and researcher for the ULPGC until 2015, when he moved to the Royal Veterinary College, UK, to undertake a 3-year clinical residency in anatomic pathology. Currently, he is a lecturer in comparative pathology at the Royal Veterinary College and consultant histopathologist at the Francis Crick Institute, London. He is a Diplomate of the American College of Veterinary Pathologists (Anatomic Pathology, is ABVS®, ACVP), an RCVS Recognized Specialist in Veterinary Pathology, holds a Master's degree in veterinary medicine, and holds a PhD and a Postdoc Fellowship in comparative oncology. He is a regular reviewer of several medical peer-reviewed journals and a council member of the Spanish Society of Veterinary Pathology. He has authored and co-authored over 80 peer-review publications covering various aspects of pathology and comparative pathology.

Gustavo A. Ramírez

Gustavo A. Ramírez graduated as a Doctor in Veterinary Medicine (DVM) in 1999 from the University of Las Palmas de Gran Canaria, Spain, and received his PhD in animal health and comparative pathology after writing a thesis on the neuroendocrine cutaneous cells (Merkel cells) and their dysfunctions in dogs. He moved to Barcelona, Spain, working on diagnostics and the characterization of neoplastic diseases in domestic animals as a researcher and veterinary pathologist from 2006 to 2017. He moved to Lleida, Spain, in 2017 and became a senior lecturer in veterinary pathology and the head of the Pathology Unit at the Veterinary College of the University of Lleida. He also coordinates the Veterinary Pathology Diagnostic and Research Lab of this university. His research has included different aspects in veterinary pathology, including the study of the cutaneous diffuse neuroendocrine system, the characterization and diagnosis of cutaneous and neoplastic disorders, the study of prognosis factors in neoplastic diseases, and the development of innovative vaccines and diagnostics for endemic and zoonotic reproductive pathogens of livestock. Dr. Ramírez is an EBVS® European Specialist in Veterinary Anatomic Pathology (Diplomate of the European College of Veterinary Pathologist) and has been the president of the Spanish Veterinary Anatomic Pathology Society since 2019.

Preface to "Comparative Pathology and Immunohistochemistry of Veterinary Species"

Comparative pathology plays a fundamental role in the One Health concept and has the ultimate goal of advancing our medical knowledge of both human and veterinary species. Immunohistochemistry is a routine diagnostic technique that is often essential to reach a final diagnosis. It is usually part of any anatomic pathology study in cancer, infectious, or developmental diseases. There has been an exponential increase in use of immunohistochemical techniques in the last few decades, reflecting the current position that immunohistochemistry holds in most pathological laboratories and research centres. This book represents a collection of papers that aim to highlight the importance of immunohistochemical findings and their role in improving the understanding of pathogenesis and diagnosis of disease in domestic and wild animals, with a special focus on comparative aspects with the human species. It is mainly directed at veterinary students and professionals, researchers, and technicians who want to improve their immunohistochemical skills and update their knowledge in pathology, microbiology, and infectious diseases in both domestic and wildlife species.

Alejandro Suárez-Bonnet and Gustavo A. Ramírez
Editors

 **animals**

Editorial

Veterinary Comparative Pathology, a Scientific Tool for a Thriving Planet

Alejandro Suárez-Bonnet [1],* and Gustavo A. Ramírez Rivero [2]

[1] The Royal Veterinary College, University of London, London WC1E 7HU, UK
[2] Agrifood, Forestry and Veterinary Campus, Universitat de Lleida, 25003 Lleida, Spain
* Correspondence: asuarezbonnet@rvc.ac.uk

In recent years, Earth has overcome unpredictable challenges. From drastic climate change to a viral pandemic of probable zoonotic origin, these concurrent events have raised, to an unprecedented scale, the awareness that the natural world dominates humankind and not vice versa.

Where does veterinary pathology fit in this storm of events? More than a century ago, Virchow Robins, the son of a butcher and often seen as the founder of modern medicine and pathology, and other contemporary scientists like Robert Koch and Louis Pasteur, worked with an enviable mind on diseases that affected both animals and humans. William Osler, a human physician, also studied parasitic diseases in animals and named the parasite *Oslerus osleri* a nematode of canids. In fact, we owe the term 'One Medicine', the precursor of 'One Health', to Dr. Osler. As science used to be more open and scientists considered the natural world, human beings, and nonhuman animals as one, the concept of 'One Medicine' may have naturally developed. In contrast, current medical, biomedical, and basic research works focus on their individual areas, overlooking the bigger picture of the 'One Health' concept.

In this Special Issue, now edited as a book entitled *Comparative Pathology and Immunohistochemistry of Veterinary Species*, the reader can find pathology-focused, rigorously peer-reviewed manuscripts on different aspects of veterinary pathology from a comparative pathology perspective with a particular focus on immunohistochemistry as an ancillary diagnostic tool and a complementary technique in pathology research.

As veterinary pathologists, we have broad knowledge of disease processes in various species with variable physiology and response to disease. The variety of areas where comparative veterinary pathology will elucidate the pathogenesis of vertebrate and invertebrate animal diseases is as wide as the number of diseases, aetiologies, and species known by science, whether they are neoplastic, infectious, non-infectious, or environmentally related.

Comparative pathology, or as we prefer to call it, 'One Pathology', is the cornerstone of biomedical sciences and, ultimately, veterinary and human medicine. Only if we understand that there are no physical or philosophical boundaries between species, diseases, and the natural world and that they are, in reality, related and causally linked will we be able to unveil the secrets to a healthier planet.

Conflicts of Interest: The authors declare no conflict of interest.

Disclaimer/Publisher's Note: The statements, opinions and data contained in all publications are solely those of the individual author(s) and contributor(s) and not of MDPI and/or the editor(s). MDPI and/or the editor(s) disclaim responsibility for any injury to people or property resulting from any ideas, methods, instructions or products referred to in the content.

Citation: Suárez-Bonnet, A.; Ramírez Rivero, G.A. Veterinary Comparative Pathology, a Scientific Tool for a Thriving Planet. *Animals* **2023**, *13*, 1504. https://doi.org/10.3390/ani13091504

Received: 5 April 2023
Revised: 19 April 2023
Accepted: 26 April 2023
Published: 28 April 2023

Article

KIT Somatic Mutations and Immunohistochemical Expression in Canine Oral Melanoma

Ginevra Brocca [1,*], Beatrice Poncina [1], Alessandro Sammarco [1,2], Laura Cavicchioli [1] and Massimo Castagnaro [1]

[1] Department of Comparative Biomedicine and Food Science, University of Padua, Legnaro, 35020 Padua, Italy; beatriceponcina@gmail.com (B.P.); alessandro.sammarco@unipd.it (A.S.); laura.cavicchioli@unipd.it (L.C.); massimo.castagnaro@unipd.it (M.C.)

[2] Department of Neurology and Radiology, Massachusetts General Hospital, Harvard Medical School, Boston, MA 02129, USA

* Correspondence: ginevra.brocca@gmail.com

Received: 9 November 2020; Accepted: 7 December 2020; Published: 10 December 2020

Simple Summary: Malignant melanomas arising from mucosal sites are very aggressive neoplastic entities which affect both humans and dogs. The family of tyrosine kinase receptors has been increasingly studied in humans for this type of neoplasm, especially the gene coding for the proto-oncogene *KIT*, and tyrosine kinase inhibitors are actually available as treatment. However, *KIT* alteration status in canine oral melanoma still lacks characterization. In this study, we investigated the mutational status and the tissue expression of *KIT* through DNA sequencing and immunohistochemical analysis, respectively. A homogeneous cohort of 14 canine oral melanomas has been collected, and while tissue expression of the protein was detected, no mutations were identifiable, most likely attributing the dysregulation of this oncogene to a more complex pattern of genomic aberration.

Abstract: Canine oral melanoma (COM) is an aggressive neoplasm with a low response to therapies, sharing similarities with human mucosal melanomas. In the latter, significant alterations of the proto-oncogene *KIT* have been shown, while in COMs only its exon 11 has been adequately investigated. In this study, 14 formalin-fixed, paraffin-embedded COMs were selected considering the following inclusion criteria: unequivocal diagnosis, presence of healthy tissue, and a known amplification status of the gene *KIT* (seven samples affected and seven non-affected by amplification). The DNA was extracted and *KIT* target exons 13, 17, and 18 were amplified by PCR and sequenced. Immunohistochemistry (IHC) for KIT and Ki67 was performed, and a quantitative index was calculated for each protein. PCR amplification and sequencing was successful in 97.62% of cases, and no single nucleotide polymorphism (SNP) was detected in any of the exons examined, similarly to exon 11 in other studies. The immunolabeling of KIT was positive in 84.6% of the samples with a mean value of 3.1 cells in positive cases, yet there was no correlation with aberration status. Our findings confirm the hypothesis that SNPs are not a frequent event in *KIT* activation in COMs, with the pathway activation relying mainly on amplification.

Keywords: Canine oral melanoma (COM); copy number aberration (CNA); dog; immunohistochemistry (IHC); single nucleotide polymorphism (SNP); receptor tyrosine kinase (KIT)

---

1. Introduction

The proto-oncogene *KIT*, firstly identified as a homolog of the feline sarcoma viral oncogene *v-kit* [1], encodes for a type 3 receptor tyrosine kinase (KIT) and it is expressed in a wide range of healthy cells [2,3]. Even if its precise role is not completely understood [4], *KIT* is also normally expressed in the development of melanocytes [2] and it is involved in melanogenesis and melanocyte survival during migration from the neural crest to the cutis. It appears to be more involved in melanocyte migration rather than proliferation [5] with interesting implications since *KIT* mutant melanocytes could acquire elevated migration abilities. Indeed, given its key role, the gene coding for the KIT protein (*KIT*) received great attention in the field of human oncology, particularly in the study of neoplasms of melanocytic origin. The genomes of both human cutaneous and mucosal melanomas were deeply studied, and human mucosal melanomas (hMMs) showed a higher frequency of aberrations in the *KIT* gene rather than their cutaneous counterparts [3,6–9]. HMMs are affected by both structural alterations, such as copy number aberrations (CNAs), and single nucleotide polymorphisms (SNPs) of *KIT*, with CNAs in 26.3% of cases [6,8] and SNPs in 7–23% [4,6–13]. Therefore, *KIT* mutations have been proposed as an adverse prognostic factor [13]. On the contrary, in cutaneous melanomas, *KIT* is affected by CNAs in 6.7% of cases and by SNPs in 1.7% of cases [8].

Mutations identified in hMMs mainly affected exon 11 (Ex-11) [8,9], where four hotspots have been identified as driver mutations [7].

In humans, most tumors affected by *KIT* mutations [3,14] mainly show SNPs affecting Ex-11 (65% of GastroIntestinal Stromal Tumors, GISTs [3]), promoting the constitutive activation of the KIT receptor without binding to the specific ligand, and leading to uncontrolled proliferation and survival of neoplastic cells [14].

Given the increasing interest in pet dogs as a reliable spontaneous animal model for the study of non-UV-induced melanomas, several authors are also investigating canine melanomas arising in sun-protected sites such as canine oral melanoma (COM), the most frequent neoplasia of the oral cavity in dogs [15–17]. Several studies noted some similarities in can profiles between hMMs [10,18–21] and COMs [10,22–24]. In a recent comparative investigation, we were able to detect common chromosomal changes in 32 regions affecting human chromosomes (HSA) and the canine orthologous regions (CFA), with amplification in 35% of cases of the KIT-coding region located on CFA 13 [25].

However, regarding the SNP profile of COMs, the majority of the studies focused almost entirely on Ex-11, trying to compare the results obtained in human medicine without investigating other exons that play a key role in *KIT* activation in hMMs, such as exons 13 (Ex-13), 17 (Ex-17) and 18 (Ex-18) [6–8,13,26,27].

When considering the mutations reported in the available human literature, Ex-11 is altered in 67% of hMMs [6–9,12,13,20,28–32], while Ex-13 is affected in 15.2% [4,6–9,12,13,20,29–32], Ex-17 in 12.2% [4,6–9,13,20,29–32], and Ex-18 in 8.5% [6,7,13,29,30,32].

In veterinary medicine, only mutations affecting Ex-11 have been adequately investigated in the *KIT* gene, but no decisive results have been achieved, while a screening of Ex-13, Ex-17, and Ex-18 has not been performed [33,34].

The tissue expression of the KIT protein has also been investigated by immunohistochemistry (IHC) in various neoplasms, both in human and veterinary medicine (the latter particularly in pet dogs [35–38]). Although less documented in COMs, and with different ranges of intensities, 49–51% of neoplasms analyzed expressed the protein [34,39] in contrast to hMMs, where 74–89% of cases were reported to be positive [9,12,26].

These heterogeneous results, both in human and veterinary studies, also bear consequences in the use and effectiveness of therapeutic approaches to *KIT*-bearing mutation tumors.

In humans, the use of tyrosine kinase inhibitors (TKIs) has a better effect on patients affected by tumors bearing SNPs in *KIT* [8,13,27,40], particularly in GIST [41], some types of melanomas [11,20,40], chronic myeloid leukemia [42], and systemic mastocytosis [43].

Moreover, the disease control rate in human patients treated with imatinib, one of the most widely used TKIs, is better in patients bearing tumors with *KIT* point mutations (77%) compared to patients bearing tumors with *KIT* amplification (18%) [40], particularly when SNPs affect Ex-11 and Ex-13 [8,13,27]. For patients with SNPs in Ex-17 and Ex-18, which are not responsive to imatinib, responsiveness to treatment with MEK-1 inhibitors has instead been suggested [8].

TKIs are also successfully used in veterinary medicine for the treatment of Mast Cell Tumors (MCTs) with *KIT* point mutations in Ex-11 [34,44], and occasionally of GIST [45,46]. Moreover, although imatinib appears to be effective mostly for tumors with *KIT* point mutations, disease regression in dogs and cats treated with imatinib and affected by tumors without known *KIT* point mutations is rare [47].

In this study, we investigated the role of *KIT* in COMs at both the gene and protein level. Taking advantage of the cohort of COMs assembled for the array Comparative Genomic Hybridization (aCGH) study published in 2019 [25], we decided to deepen our knowledge of the possible role of the gene *KIT* and the protein KIT in our cohort of samples. We performed an accurate evaluation of the possible SNPs by sequencing Ex-13, Ex-17, and Ex-18, and by evaluating KIT expression in COMs. Ki67 was also evaluated by IHC due to its higher expression in metastatic hMMs with SNPs in *KIT* [13].

Therefore, the aim of this study was to evaluate and correlate the IHC expression of KIT and Ki67 with *KIT* somatic point mutations and amplification in COMs.

2. Materials and Methods

2.1. First Case Selection

Formalin-fixed, paraffin-embedded (FFPE) COMs were collected from different archives. For each patient, anamnestic data including size and location of the tumor were collected. Hematoxylin and eosin (H&E)-stained slides from each sample were checked by 3 experienced pathologists to confirm the diagnosis of COM and to assess the presence of healthy tissue near the pathological mass suitable for the nucleic acid extraction. The diagnosis of COM was made by adopting the criteria proposed by Smedley and colleagues [48], which were also used for pigmentation scoring, and uncertain cases (i.e., amelanotic specimens) were tested via IHC with anti-PNL-2 and anti-Melan-A antibodies. Only cases that were unanimously diagnosed as COM by all the reviewing pathologists were taken into consideration. At the end of phase one, forty cases met all the inclusion criteria and were selected for the following analyses.

2.2. Nucleic Acid Extraction

With the help of a microtome with disposable blades, 2–3 20 µm-thick slides were cut from each FFPE block. Healthy and cancerous tissues were separated using a clean scalpel and put in 2 different tubes. To avoid any possible contamination, all the instruments were cleaned between one sample and the other. From the paraffinized material, DNA was extracted using the All Prep DNA/RNA Extraction Kit (Qiagen, Hilden, Germany®®) using heptane as the deparaffinization agent and following the manufacturer's instructions. The extracted DNA was then evaluated on a 1% agarose gel by electrophoresis and quantified with the NanoDrop ND-1000 (Thermo Fisher Scientific, Waltham, MA, USA®®).

2.3. aCGH Analysis and Second Case Selection

Twenty of the 40 samples showed an adequate DNA yield and were submitted to an aCGH analysis as described in the work from Brocca and colleagues [25]. The aCGH allowed us to learn the aberrational status of CFA 13, and in particular of the portion containing the *KIT* gene. All the samples affected by a copy number gain in the *KIT* locus were selected for the study, together with an equal number of samples not affected by a copy number gain involving *KIT*. The latter were randomly chosen among the remaining samples submitted to the aCGH analysis to maintain a balanced number of samples between the 2 groups (Table 1).

Table 1. Selected samples included in the study, together with available clinical and histological data, immunohistochemistry (IHC) indexes, and amplification status. B: bad prognosis; F: female; FN: neutered female; G: good prognosis; M: male; MN: neutered male; NA: not available; * mean number of positive cells in 5 high-power fields (hpf).

Case ID	Site (Oral Cavity)	Breed	Age (Years)	Sex	Pigmentation [48]	*KIT* locus Amplification	Prognosis [49]	Ki67 Index * [49]	KIT Index *
1	Mandible	West Highland White Terrier	11	M	<50%	yes	G	16.2	8.6
2	Cheek	Cross breed	17	M	<50%	yes	B	137.4	0.8
3	Oral	Rottweiler	12	FN	<50%	no	B	29.6	1.4
4	Lip	American Cocker Spaniel	10	MN	<50%	yes	B	54.2	5
5	Upper lip	Golden Retriever	12	M	<50%	yes	B	113.4	6.6
6	Lip	Cocker Spaniel	10	F	<50%	no	B	26.2	6.6
7	Mandible	Pug	11	M	<50%	yes	B	37.4	NA
8	Oral	Collie	11	M	≥50%	yes	B	30	0
9	Upper lip	German Shepherd	11	M	<50%	yes	B	22.6	0.6
10	Mandible	Cocker Spaniel	11	F	≥50%	no	B	21.6	2
11	Upper lip	Pinscher	17	M	≥50%	no	G	7.6	3
12	Upper lip	Cocker Spaniel	10	M	≥50%	no	B	265	3.8
13	Upper lip	Basset Hound	11	M	<50%	no	G	9.8	0
14	Upper lip	Golden Retriever	13	M	<50%	no	B	102.4	1.6
							Mean	62.4	3.1

2.4. Exon Amplification and Sequencing

Primers were designed (Table 2) to amplify the exons considered in this study (Ex-13, Ex-17, and Ex-18). For each exon, reference sequences (CanFam3.1 annotation) were obtained from the online platform Ensembl [50], and primers were designed with the software Primer3 v4.1 [51]. Amplicon size and coding sequence covered for each exon are reported in Table 2.

Table 2. Primers designed for the amplification of exons 13, 17, and 18 of the *KIT* gene with the relative lengths. CDS: Coding DNA Sequence; F: forward; R: reverse.

		Primer Sequence	Primer Length	Annealing Temp	Primer Genomic Location	Amplicon Size (bp)	CDS Covered (%)
Exon 13	F Primer	TGGCTTGCCAAATTTGC TTCT	21	59.58° C	13:47108547–47108567	247 bp	100%
	R Primer	AACCAAGCACTGTCGC AATG	20	59.69 °C	13:47108774–47108793		
Exon 17	F Primer	TGACATAGCAGCATTCTC GTGT	22	60.09 °C	13:47113635–47113656	257 bp	100%
	R Primer	TCCTTCACTGGACTGTC AAGC	21	59.93 °C	13:47113871–47113891		
Exon 18	R Primer	CATTGCCGGATCTGTT GTGC	20	60.18 °C	13:47117315–47117334	211 bp	100%
	F Primer	AGGACCCTGCTAACCC CTTA	20	59.58 °C	13:47117506–47117525		

A polymerase chain reaction was then performed for all samples with an initial denaturation step for 2 min at 94 °C, 42 cycles of 40 s at 94 °C (denaturing), 40 s at 60.5 °C (annealing), 50 s at 72 °C (extension), and 5 min of final extension at 72 °C.

The positive control was non-fragmented genomic DNA extracted from fresh-frozen canine cutis, while water was used instead of DNA as the negative control.

All amplicons obtained were separated on a 1.8% agarose gel via electrophoresis to assess the success of the PCR reaction, then purified with ExoSAP (Exonuclease I Shrimp Alkaline Phosphatase, Thermo Fisher Scientific®®). Using the corresponding forward (Ex-17) or reverse (Ex-13 and Ex-18) primer, amplicons were then sequenced with the Sanger method, and their sequences were visualized with the Chromas 2.6.5 software. Only sequences with high-resolution peaks (high signal-to-noise ratio), minimal baseline noise, and no trace of secondary sequence contamination were considered suitable for mutational analysis. For each sample, healthy and pathological sequences were matched and aligned with the Clustalw platform [52]. The reference sequence was used to identify SNP positions. The use of pathological and healthy tissue from the same dog allowed for the discrimination of germline and somatic SNPs, and only the latter were taken into consideration for further analyses. The presence of a specific mutation in at least 15% of the samples was arbitrarily considered the minimum threshold.

2.5. Immunohistochemistry and Immunohistochemical Assessment

KIT expression was evaluated in each sample included in this study. For each FFPE block, a 4 μm-thick slide was cut, mounted on a polarized glass slide (TOMO®®, Matsunami Glass) and then tested with an anti-KIT rabbit polyclonal antibody diluted 1:300 (Dako®®, CD117 clone).

For each specimen, the expression of the Ki67 protein was also assessed with an anti-Ki67 mouse monoclonal antibody diluted 1:50 (Dako®®, MIB1 clone). Both antibodies were previously validated in the canine species [35,53].

Both procedures were performed with an automatic immunostainer (Ventana Benchmark GX, Roche-Diagnostic). To avoid the use of a bleaching reaction which could damage the integrity of the antigens, an ultraView universal alkaline phosphatase RED detection kit (Ventana Medical System Inc., Oro Valley, AZ, USA) was used (DAB chromogen in unbleached specimens is indeed not usable), and hematoxylin was used as a counterstain.

As positive controls, a canine MCT and canine cutis were used for KIT and Ki67 staining, respectively. As negative controls, antibody diluent was applied instead of the antibody.

The slides were visualized at 40× magnification using a D-Sight scanning machine and the D-Sight Viewer software (A. Menarini Diagnostics). The KIT index was evaluated by counting the mean number of positive cells in 5 consecutive high-power fields (hpf; 0.237 mm^2) within the areas with clear positive staining, starting from the mostly positive field. If no positive cells were found in 3 consecutive hpf, the process was repeated in another IHC-positive area. This method was designed similarly to that used for the establishment of the Ki67 index as described by Bergin and colleagues [49], which was applied for the Ki67 index calculation. When more than one biopsy was present for each tumor (e.g., for margin evaluation), 5 hpf were selected for each specimen. Only areas with a cellular population representative of the tumor were selected, and areas affected by background, degeneration, scirrhous reaction, or necrosis were avoided. Neoplastic cells were considered positively KIT-labeled when they showed brightly red cytoplasmic and membrane staining, as exemplified in Figure 1A from the MCT control, and as described in the literature [35].

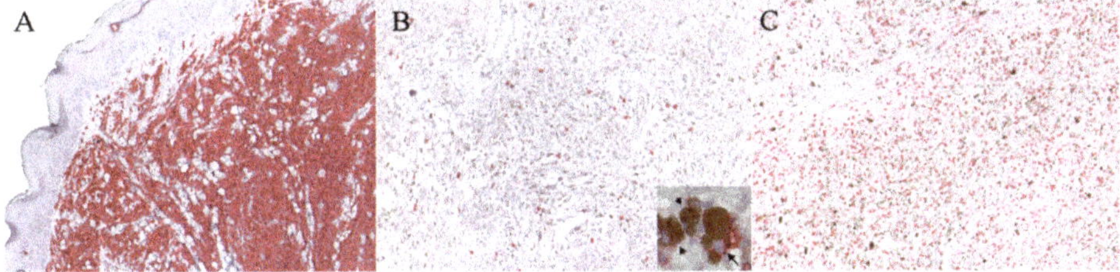

Figure 1. IHC for KIT and Ki67 with the RED labeling system. (**A**) Mast cell tumor (MCT). Example of the mast cell tumor used as the control tissue for the KIT immunolabeling, with a high density of positive cells (100× magnification). (**B,C**) Canine oral melanoma (COM). (**B**) Example of a neoplastic area selected for the evaluation of the KIT index, which was 6.6 (100× magnification): the positive cells are discernible and scattered throughout the field. The details of the positive neoplastic melanocytes (arrow) are provided with higher magnification in the inset; melanomacrophages are also present within the inset (arrowheads). (**C**) Example of a neoplastic area selected for the evaluation of the Ki67 index, which was 30 (100× magnification).

2.6. Statistical Analysis

Statistical analyses were performed using MedCalc (MedCalc Statistical Software version 15.8). Data distribution was visually checked for normality. To verify mean differences among groups, either the Student's *t* test or the one-way ANOVA with Tukey's multiple comparisons were performed when data were normally distributed. The Mann–Whitney test or Kruskal–Wallis test was applied when data were not normally distributed. A Chi-square test and Fisher's exact test were used for analysis of the association between *KIT* amplification status and clinical features.

The Spearman's rank correlation analysis was applied to discover associations between variables. The level of significance was set at $p < 0.05$.

3. Results

3.1. aCGH Analysis

Data related to the aCGH analysis are published and extensively discussed in the paper from Brocca and colleagues [25]. Twenty samples were submitted for aCGH analysis, and a total number of 14 samples were selected for this study: seven harboring a KIT amplification, and seven without KIT amplification, which were randomly chosen among the remaining 13 samples to maintain a balanced number of samples between the two groups.

3.2. Epidemiological Data

The mean age of the 14 dogs selected for the study was 11.98, ranging from 10–17 years. Males represented 78.57% (11/14, one neutered) of the cases, while females represented the remaining 21.43% (3/14, one neutered). Male over-representation is in accordance with the literature [15,54]. The most represented breed was the cocker spaniel, with 28.57% (4/14), and the most common sites of the neoplasia (when specified) was the oral lip, with 57.14% (8/14, six of which on the upper lip). For the immunohistochemical evaluation of Ki67, the majority of the samples (85.71%, 12/14) showed an index >19.5, significant of a bad prognosis at one-year post-diagnosis [49]. This data is visible in Table 1. All samples came from surgical excision or incisional biopsy.

3.3. Identification of Somatic Mutations

The DNA from all 14 samples was successfully extracted from the fractions of healthy and pathological tissues, with the use of heptane as the deparaffinizing agent to obtain a higher yield of extracted nucleic acids. From Nanodrop analysis, healthy tissues showed a lower amount of extracted DNA ($p < 0.0001$, data not shown), and the 260/280 and 260/230 ratios showed variable values, but all were considered of sufficient quality for the amplification steps. All primer pairs were firstly tested on canine control tissues at different temperatures with a gradient PCR, and all pairs showed successful and specific amplification of the selected exons. The amplification was successful in almost all samples for both the healthy and the pathologic tissues. A band at the expected length was obtained in 27/28 (96.4%) reactions (14 DNA samples from healthy tissues and 14 DNA samples from pathologic tissues) for exon 13, 27/28 (96.4%) for exon 17, and 28/28 (100%) for exon 18 (Table 3). Indeed, it was not possible to obtain an amplification product for exons 13 or 17 using the pathologic DNA extracted from sample 12 (Figure 2). All amplicons obtained were successfully sequenced and analyzed as described by three operators. In summary, 97.6% of the exonic sequences (82 sequences out of 84) examined were successfully amplified and Sanger-sequenced, and no somatic SNPs, insertions, or deletions were identified in any of the samples.

Table 3. Result of the PCR amplification obtained for each DNA fraction (healthy and pathologic) of the samples analyzed, and for each primer pair. H: healthy; P: pathologic. Ex: exon.

Case ID	DNA	Ex-13	Ex-17	Ex-18
1	H	yes	yes	yes
	P	yes	yes	yes
2	H	yes	yes	yes
	P	yes	yes	yes
3	H	yes	yes	yes
	P	yes	yes	yes
4	H	yes	yes	yes
	P	yes	yes	yes
5	H	yes	yes	yes
	P	yes	yes	yes
6	H	yes	yes	yes
	P	yes	yes	yes
7	H	yes	yes	yes
	P	yes	yes	yes
8	H	yes	yes	yes
	P	yes	yes	yes
9	H	yes	yes	yes
	P	yes	yes	yes
10	H	yes	yes	yes
	P	yes	yes	yes
11	H	yes	yes	yes
	P	yes	yes	yes
12	H	yes	yes	yes
	P	no	no	yes
13	H	yes	yes	yes
	P	yes	yes	yes
14	H	yes	yes	yes
	P	yes	yes	yes

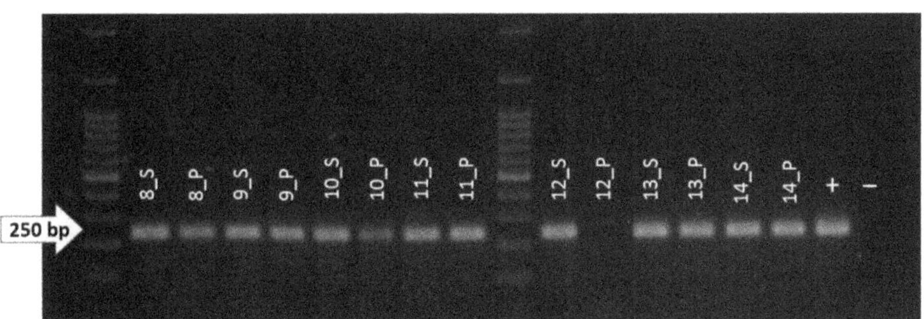

Figure 2. 1.8% agarose gel showing the amplicons obtained by PCR reaction for exon 13 (Ex-13) of the KIT gene. As shown, no amplified DNA was obtained from the pathological fraction of sample 12 (12_P). +: positive control (non-fragmented canine genomic DNA); −: negative control (water).

3.4. Immunohistochemistry

One of the samples (sample 7 in Table 1) could not be evaluated for KIT expression since the DNA extraction procedure exhausted the paraffin block.

Positive immunolabeling was obtained in the MCT control (Figure 1A) and in all the cutaneous melanocytes (internal control) of the healthy tissues included along with COM specimens.

As expected, the anti-KIT antibody showed cytoplasmic and membranous labeling of neoplastic melanocytes with a bright multifocal positivity of a scarce to a moderate number of cells (Figure 1B).

The positive cells were discernible and randomly distributed across the tumor tissue in the majority of samples; particular patterns of distribution were not noted. Only two samples (15.4%) did not show any staining for KIT expression.

The KIT index varied from 0.6 to 8.6 in positive samples, with a mean of 3.1 positive cells in 5 hpf. Positive staining for the anti-Ki67 antibody (Figure 1C) was obtained for all specimens and the Ki67 index was established as described [49]. The Ki67 index varied from 7.6 to 265, with a mean of 62.4 positive cells in 5 hpf. According to the Ki67 prognostic cut-off of 19.5, three samples were classified as having a good prognosis (G), and the remaining 11 as having a bad (B) prognosis. The results are summarized in Table 1. When samples were divided into two groups according to the KIT amplification status, no statistically significant differences were detected either for the KIT index ($p = 0.56$) or for the Ki67 index ($p = 0.38$). Similarly, no differences in terms of KIT and Ki67 expression were noted between males and females ($p = 0.87$ and $p = 0.46$, respectively), or according to the pigmentation level ($p = 0.48$ and $p = 0.73$, respectively). Finally, no correlation was found between the two indexes ($r = 0.074$, $p = 0.81$).

4. Discussion

The aim of this study was to deepen our understanding of the mutational landscape of the *KIT* gene in COMs, particularly in exons 13, 17, and 18, and to correlate the mutational profile of these exons with the amplification status of the gene itself, and with the IHC expression of the KIT protein.

This interest derives from the scarcity of currently available similar studies in the canine species, and from the possible use of pet dogs as a reliable model for the study of hMMs.

In our previous work [25] that aimed to improve our knowledge about the genomic DNA alterations that occur in COM, many genes related to MAPK and PI3K pathways were detected from the CNA analysis, together with a wide variety of genes coding for tyrosine kinases receptors, including *KIT*. Interestingly, the pathway enrichment analysis revealed the enhancement of pathways specifically related to cancer proliferation, but also a significant enrichment of those related to imatinib and drug metabolism. These results indicated that further investigation of the *KIT* alteration status was warranted.

In this work, we describe the first characterization of the mutational profile of exons 13, 17, and 18 of the *KIT* gene in COM, which were successfully PCR amplified and Sanger-sequenced. Taking advantage of the cohort of samples collected for the aCGH study [25], we were able to compare the exon sequences of healthy and pathologic tissues from COMs with known *KIT* amplification status. In particular, CFA 13 (comprising the *KIT* gene) was affected by a copy number gain in 7/20 samples of the original cohort (35%), and we considered it valuable to further analyze the DNA of these seven samples and to compare them with another randomly chosen seven samples that were not affected by the same copy number gain in CFA 13.

We developed highly-performing primer pairs and set up a reliable protocol for the amplification and sequencing of short genomic sequences extracted from FFPE blocks.

Since no SNPs were detected affecting the examined exons, our study suggests that *KIT* status in COMs does not resemble the mutational status reported in hMMs. This is in line with some of the latest Next Generation Sequencing-based veterinary studies [10,24]. Although Garrido and Bastian [4] suggested that CNAs and SNPs in hMM are mostly mutually exclusive, two different studies reported *KIT* amplification and coexisting SNPs in exons 11, 13, 17 and 18 in the same tumor [6,8].

In our study, none of the samples affected by a CFA 13 amplification had an SNP present, and no point mutations have been found at all, suggesting one possible molecular difference between COMs and hMMs.

The absence of point mutations in our cohort is consistent with the results reported in other recent studies of COMs, in which the *KIT* point mutations are considered a sporadic event [10,33,34,47], which highlights a potential significant molecular difference with hMMs [4,8,10].

The detection of a 35% prevalence of *KIT* amplification versus a 0% prevalence of *KIT* point mutations in our cohort of COMs corroborates the increasingly affirmed hypothesis that the main pathogenesis of COMs and hMMs is related predominantly to CNAs rather than SNPs [10,18–21].

Regarding the pathologic DNA from sample 12, we were not able to obtain an amplification reaction for exon 13 and 17, while the amplification of exon 18 was successful (Table 3). The reason for this could be the high melanin content of the sample. Indeed, melanin is an interferer of the PCR reaction and other molecular analysis when it is co-purified in the process of DNA extraction. Moreover, it has already been demonstrated that PCRs producing longer amplicons are more inclined to be inhibited by melanin than PCRs producing amplicons of shorter size [55]. In our case, melanin could have bound to the DNA polymerase enzyme, preventing the PCR reaction in the longer exons, i.e., exons 13 and 17, which both had a length close to 250 bp, but not in the shorter exon 18, which is approximately 200 bp long.

Here, we proposed a reproducible method for scoring KIT IHC positivity in COMs samples. To date, the IHC evaluation of KIT expression has been limited to a semi-quantitative evaluation, expressed as classes corresponding to an approximate percentage of immunoreactive cells on the total tumor area, often with wide ranges defining a single class [34,39]. In our opinion, this approach poorly describes the mutable status of expression of the protein. In support of this hypothesis and in contrast with other studies, our IHC results closely reflect the human literature: 84.6% of our samples were indeed positive, and therefore nearer to the percentage provided in human literature. In other veterinary studies, only about half of the COMs examined (49–51%) [34,39] were considered positive with the semi-quantitative scoring method. It is still unclear if this difference is related to the different scoring method or to a real difference in KIT expression between hMMs and COMs, and further evaluation (or a re-evaluation of previous works) is necessary. We also semi-quantitatively scored our samples following the methods proposed in [34,39], and we noted an overestimation of negative cases when the percentage of positive cells was <10% (data not shown).

There was no significant difference in KIT protein expression between *KIT* amplified and non-amplified samples. This could be due to the low number of cases analyzed, or to the fact that *KIT* gene amplification does not correspond to a higher KIT protein expression.

As reported by Lassam and Bickford [56], and by Montone and colleagues [57], an interesting observation regarding KIT in melanomas is the decrease (or even the loss) of KIT expression along with the progression of the neoplastic disease. This was observed in human cultured melanoma cells [56] and in cutaneous melanomas (from radial growth phase to vertical growth phase and metastatic melanoma) [57], which led to the hypothesis that the loss of KIT could represent a negative prognostic factor [56,57].

These findings support the hypothesis of Alexeev and Yoon [5], who proposed that for malignant melanocytes to acquire metastatic potential and escape from the epidermal boundaries [4], they necessarily have to lose KIT expression.

Indeed, a study from Newman et al. [39] found a significant association between the presence of KIT IHC positivity and patient survival, suggesting that the downregulation or loss of KIT could be related to increasing invasiveness in dogs as well.

Regarding the correlation with other tumor markers, Ma and colleagues [13] highlighted an increased Ki67 expression in metastatic hMM, while the lack of pigment was considered a negative prognostic factor by Prouteau and colleagues [58]. However, in our study, no statistically significant difference was noted in either case.

Unfortunately, follow-up data were not available for this cohort of samples, making further evaluations impossible.

5. Conclusions

In this paper, we showed that exons 13, 17, and 18 of the *KIT* gene do not present with SNPs in our cohort of COMs, suggesting that they are not involved in the pathogenesis or progression of COM.

Moreover, the amplified status was not statistically associated in any way to a mutational profile in the *KIT* gene, nor to the expression of the KIT protein itself or to the Ki67 marker.

Immunohistochemically, we proposed a quantitative, replicable method for the evaluation of KIT expression, through which 84.6% of the samples (11/13) were detected to be positive for KIT. Although the number of cases included in this study did not allow us to draw definitive conclusions, our findings provide new insights for the current knowledge on the use of pet dogs as a spontaneous model for the study of hMMs. Further studies with a greater number of cases are needed to clarify unresolved questions, including the role played by other kinases that have also been found to be altered in our cohort via aCGH analysis, such as *PTK2, STK3, TEC, PDGFRA, VEGFR2,* and *CD63,* which so far have received less attention.

Author Contributions: Conceptualization, G.B. and M.C.; methodology, G.B. and M.C.; formal analysis, G.B. and A.S.; investigation, B.P. and G.B.; resources, L.C. and M.C.; writing—original draft preparation, B.P and G.B.; writing—review and editing, A.S., L.C. and M.C.; visualization, G.B. and A.S.; supervision, M.C.; project administration, M.C.; funding acquisition, L.C. and M.C. All authors have read and agreed to the published version of the manuscript.

Funding: This research received no external funding.

Acknowledgments: The authors would like to thank Iussich (from the Department of Veterinary Science of the University of Turin) and Peña (from the Department of Animal Medicine, Surgery and Pathology of the Complutense University of Madrid) for providing some of the FFPE samples. The authors also thank Ferraresso, Zanetti and Guerra (from the Department of Comparative Biomedicine and Food Science of the University of Padua) for their support with laboratory practices. Sammarco is supported by an American–Italian Cancer Foundation Post-Doctoral Research Fellowship.

Conflicts of Interest: The authors declare no conflict of interest.

References

1. Yarden, Y.; Kuang, W.J.; Yang-Feng, T.; Coussens, L.; Munemitsu, S.; Dull, T.J.; Chen, E.; Schlessinger, J.; Francke, U.; Ullrich, A. Human proto-oncogene c-kit: A new cell surface receptor tyrosine kinase for an unidentified ligand. *Embo J.* **1987**, *6*, 3341–3351. [CrossRef]
2. Nishikawa, S.; Kusakabe, M.; Yoshinaga, K.; Ogawa, M.; Hayashi, S.; Kunisada, T.; Era, T.; Sakakura, T.; Nishikawa, S. In utero manipulation of coat color formation by a monoclonal anti-c-kit antibody: Two distinct waves of c-kit-dependency during melanocyte development. *Embo J.* **1991**, *10*, 2111–2118. [CrossRef]
3. Babaei, M.A.; Kamalidehghan, B.; Mohammad, S.; Huri, H.; Ahmadipour, F. Receptor tyrosine kinase (c-Kit) inhibitors: A potential therapeutic target in cancer cells. *Drug Des. Devel.* **2016**, *10*, 2443–2459. [CrossRef] [PubMed]
4. Garrido, M.; Bastian, B.C. Kit as a therapeutic target in melanoma. *J. Invest. Derm.* **2010**, *130*, 20–27. [CrossRef] [PubMed]
5. Alexeev, V.; Yoon, K. Distinctive role of the cKit receptor tyrosine kinase signaling in mammalian melanocytes. *J. Investig. Dermatol.* **2006**, *126*, 1102–1110. [CrossRef] [PubMed]
6. Curtin, J.A.; Busam, K.; Pinkel, D.; Bastian, B.C. Somatic activation of KIT in distinct subtypes of melanoma. *J. Clin. Oncol.* **2006**, *24*, 4340–4346. [CrossRef] [PubMed]
7. Dumaz, N.; André, J.; Sadoux, A.; Laugier, F.; Podgorniak, M.P.; Mourah, S.; Lebbé, C. Driver KIT mutations in melanoma cluster in four hotspots. *Melanoma Res.* **2015**, *25*, 88–90. [CrossRef]
8. Beadling, C.; Jacobson-Dunlop, E.; Hodi, F.S.; Le, C.; Warrick, A.; Patterson, J.; Town, A.; Harlow, A.; Cruz, F.; Azar, S.; et al. KIT gene mutations and copy number in melanoma subtypes. *Clin. Cancer Res.* **2008**, *14*, 6821–6828. [CrossRef]
9. Torres-cabala, C.A.; Wang, W.; Trent, J.; Yang, D.; Chen, S.; Kim, K.B.; Woodman, S.; Davies, M.; Plaza, J.A.; Nash, J.W.; et al. Correlation between KIT expression and KIT mutation in melanoma: A study of 173 cases with emphasis on the acral- lentiginous/mucosal type. *Mod. Pathol.* **2009**, *22*, 1446–1456. [CrossRef]
10. Wong, K.; van der Weyden, L.; Schott, C.R.; Foote, A.; Constantino-Casas, F.; Smith, S.; Dobson, J.M.; Murchison, E.P.; Wu, H.; Yeh, I.; et al. Cross-species genomic landscape comparison of human mucosal melanoma with canine oral and equine melanoma. *Nat. Commun.* **2019**, *10*. [CrossRef]

11. Antonescu, C.R.; Busam, K.J.; Francone, T.D.; Wong, G.C.; Guo, T.; Agaram, N.P.; Besmer, P.; Jungbluth, A.; Gimbel, M.; Chen, C.T.; et al. L576P KIT mutation in anal melanomas correlates with KIT protein expression and is sensitive to specific kinase inhibition. *Int. J. Cancer* **2007**, *121*, 257–264. [CrossRef] [PubMed]

12. Rivera, R.S.; Nagatsuka, H.; Gunduz, M.; Cengiz, B.; Gunduz, E.; Siar, C.H.; Tsujigiwa, H.; Tamamura, R.; Han, K.N.; Nagai, N. C-kit protein expression correlated with activating mutations in KIT gene in oral mucosal melanoma. *Virchows Arch.* **2008**, *452*, 27–32. [CrossRef] [PubMed]

13. Ma, X.; Wu, Y.; Zhang, T.; Song, H.; Jv, H.; Guo, W.; Ren, G. The clinical significance of c-Kit mutations in metastatic oral mucosal melanoma in China. *Oncotarget* **2017**, *8*, 82661–82673. [CrossRef] [PubMed]

14. London, C.A. Tyrosine kinase inhibitors in veterinary medicine. *Top. Companion Anim. Med.* **2009**, *24*, 106–112. [CrossRef] [PubMed]

15. Smith, S.H.; Goldschmidt, M.H.; McManus, P.M. A comparative review of melanocytic neoplasms. *Vet. Pathol.* **2002**, *39*, 651–678. [CrossRef] [PubMed]

16. Todoroff, R.J.; Brodey, R.S. Oral and pharyngeal neoplasia in the dog: A retrospective survey of 361 cases. *J. Am. Vet. Med. Assoc.* **1979**, *175*, 567–571. [PubMed]

17. Goldschmidt, M.H. Benign and malignant melanocytic neoplasms of domestic animals. *Am. J. Derm.* **1985**, *7*, 203–212. [CrossRef]

18. Curtin, J.A.; Fridlyand, J.; Kageshita, T.; Patel, H.N.; Busam, K.J.; Kutzner, H.; Cho, K.H.; Aiba, S.; Bröcker, E.B.; LeBoit, P.E.; et al. Distinct sets of genetic alterations in melanoma. *N. Engl. J. Med.* **2005**, *353*, 2135–2147. [CrossRef]

19. Furney, S.J.; Turajlic, S.; Stamp, G.; Nohadani, M.; Carlisle, A.; Thomas, J.M.; Hayes, A.; Strauss, D.; Gore, M.; Van Den Oord, J.; et al. Genome sequencing of mucosal melanomas reveals that they are driven by distinct mechanisms from cutaneous melanoma. *J. Pathol.* **2013**, *230*, 261–269. [CrossRef]

20. Lyu, J.; Song, Z.; Chen, J.; Shepard, M.J.; Song, H.; Ren, G.; Li, Z.; Guo, W.; Zhuang, Z.; Shi, Y. Whole-exome sequencing of oral mucosal melanoma reveals mutational profile and therapeutic targets. *J. Pathol.* **2018**, *244*, 358–366. [CrossRef]

21. Hayward, N.K.; Wilmott, J.S.; Waddell, N.; Johansson, P.A.; Field, M.A.; Nones, K.; Patch, A.M.; Kakavand, H.; Alexandrov, L.B.; Burke, H.; et al. Whole-genome landscapes of major melanoma subtypes. *Nature* **2017**, *545*, 175–180. [CrossRef] [PubMed]

22. Poorman, K.; Borst, L.; Moroff, S.; Roy, S.; Labelle, P.; Motsinger-Reif, A.; Breen, M. Comparative cytogenetic characterization of primary canine melanocytic lesions using array CGH and fluorescence in situ hybridization. *Chromosom. Res.* **2015**, *23*, 171–186. [CrossRef] [PubMed]

23. Hendricks, W.P.D.; Zismann, V.; Sivaprakasam, K.; Legendre, C.; Poorman, K.; Tembe, W.; Kiefer, J.; Liang, W.; DeLuca, V.; Stark, M.; et al. Somatic inactivating PTPRJ mutations and dysregulated pathways identified in canine malignant melanoma by integrated comparative genomic analysis. *PLoS Genet.* **2018**, *14*, e1007589. [CrossRef] [PubMed]

24. Giannuzzi, D.; Marconato, L.; Ramy, E.; Ferraresso, S.; Scarselli, E.; Fariselli, P.; Nicosia, A.; Pegolo, S.; Leoni, G.; Laganga, P.; et al. Longitudinal transcriptomic and genetic landscape of radiotherapy response in canine melanoma. *Vet. Comp. Oncol.* **2019**. [CrossRef] [PubMed]

25. Brocca, G.; Ferraresso, S.; Zamboni, C.; Martinez-Merlo, E.M.; Ferro, S.; Goldschmidt, M.H.; Castagnaro, M. Array Comparative Genomic Hybridization analysis reveals significantly enriched pathways in Canine Oral Melanoma. *Front. Oncol.* **2019**, *9*, 1397. [CrossRef] [PubMed]

26. Ashida, A.; Takata, M.; Murata, H.; Kido, K.; Saida, T. Pathological activation of KIT In metastatic tumors of acral and mucosal melanomas. *Int. J. Cancer* **2009**, *124*, 863–868. [CrossRef] [PubMed]

27. Wei, X.; Mao, L.; Chi, Z.; Sheng, X.; Cui, C.; Kong, Y.; Dai, J.; Wang, X.; Li, S.; Tang, B.; et al. Efficacy evaluation of imatinib for the treatment of melanoma: Evidence from a retrospective study. *Oncol. Res.* **2019**, *27*, 495–501. [CrossRef]

28. Chen, F.; Zhang, Q.; Wang, Y.; Wang, S.; Feng, S.; Qi, L.Y.; Li, X.; Ding, C. KIT, NRAS, BRAF and FMNL2 mutations in oral mucosal melanoma and a systematic review of the literature. *Oncol. Lett.* **2018**, *15*, 9786–9792. [CrossRef]

29. Yun, J.; Lee, J.; Jang, J.; Lee, E.J.; Jang, K.T.; Kim, J.H.; Kim, K.M. KIT amplification and gene mutations in acral/mucosal melanoma in Korea. *Apmis* **2011**, *119*, 330–335. [CrossRef]

30. Omholt, K.; Grafström, E.; Kanter-Lewensohn, L.; Hansson, J.; Ragnarsson-Olding, B.K. KIT pathway alterations in mucosal melanomas of the vulva and other sites. *Clin. Cancer Res.* **2011**, *17*, 3933–3942. [CrossRef]

31. Zebary, A.; Jangard, M.; Omholt, K.; Ragnarsson-Olding, B.; Hansson, J. KIT, NRAS and BRAF mutations in sinonasal mucosal melanoma: A study of 56 cases. *Br. J. Cancer* **2013**, *109*, 559–564. [CrossRef] [PubMed]

32. Satzger, I.; Schaefer, T.; Kuettler, U.; Broecker, V.; Voelker, B.; Ostertag, H.; Kapp, A.; Gutzmer, R. Analysis of c-KIT expression and KIT gene mutation in human mucosal melanomas. *Br. J. Cancer* **2008**, *99*, 2065–2069. [CrossRef] [PubMed]

33. Chu, P.Y.; Pan, S.L.; Liu, C.H.; Lee, J.; Yeh, L.S.; Liao, A.T. KIT gene exon 11 mutations in canine malignant melanoma. *Vet. J.* **2013**, *196*, 226–230. [CrossRef] [PubMed]

34. Murakami, A.; Mori, T.; Sakai, H.; Murakami, M.; Yanai, T.; Hoshino, Y.; Maruo, K. Analysis of KIT expression and KIT exon 11 mutations in canine oral malignant melanomas. *Vet. Comp. Oncol.* **2011**, *9*, 219–224. [CrossRef] [PubMed]

35. Morini, M.; Bettini, G.; Preziosi, R.; Mandrioli, L. C-kit gene product (CD117) immunoreactivity in canine and feline paraffin sections. *J. Histochem. Cytochem.* **2004**, *52*, 705–708. [CrossRef]

36. Sabattini, S.; Bettini, G. An Immunohistochemical analysis of canine haemangioma and haemangiosarcoma. *J. Comp. Pathol.* **2009**, *140*, 158–168. [CrossRef]

37. Yu, C.H.; Hwang, D.N.; Yhee, J.Y.; Kim, J.H.; Im, K.S.; Nho, W.G.; Lyoo, Y.S.; Sur, J.H. Comparative immunohistochemical characterization of canine seminomas and Sertoli cell tumors. *J. Vet. Sci.* **2009**, *10*, 1–7. [CrossRef]

38. Frost, D. Gastrointestinal stromal tumors and leiomyomas in the dog: A histopathologic, immunohistochemical, and molecular genetic study of 50 cases. *Vet. Pathol.* **2003**, *40*, 42–54. [CrossRef]

39. Newman, S.J.; Jankovsky, J.M.; Rohrbach, B.W.; LeBlanc, A.K. C-kit expression in canine mucosal melanomas. *Vet. Pathol.* **2012**, *49*, 760–765. [CrossRef]

40. Hodi, F.S.; Corless, C.L.; Giobbie-Hurder, A.; Fletcher, J.A.; Zhu, M.; Marino-Enriquez, A.; Friedlander, P.; Gonzalez, R.; Weber, J.S.; Gajewski, T.F.; et al. Imatinib for melanomas harboring mutationally activated or amplified kit arising on mucosal, acral, and chronically sun-damaged skin. *J. Clin. Oncol.* **2013**, *31*, 3182–3190. [CrossRef]

41. Heinrich, M.C.; Corless, C.L.; Demetri, G.D.; Blanke, C.D.; Von Mehren, M.; Joensuu, H.; McGreevey, L.S.; Chen, C.J.; Van Den Abbeele, A.D.; Druker, B.J.; et al. Kinase mutations and imatinib response in patients with metastatic gastrointestinal stromal tumor. *J. Clin. Oncol.* **2003**, *21*, 4342–4349. [CrossRef] [PubMed]

42. Deininger, M.; Buchdunger, E.; Druker, B.J. The development of imatinib as a therapeutic agent for chronic myeloid leukemia. *Blood* **2005**, *105*, 2640–2653. [CrossRef] [PubMed]

43. Pardanani, A. Systemic mastocytosis in adults: 2019 update on diagnosis, risk stratification and management. *Am. J. Hematol.* **2019**, *94*, 363–377. [CrossRef] [PubMed]

44. Isotani, M.; Ishida, N.; Tominaga, M.; Tamura, K.; Yagihara, H.; Ochi, S.; Kato, R.; Kobayashi, T.; Fujita, M.; Fujino, Y.; et al. Effect of tyrosine kinase inhibition by imatinib mesylate on mast cell tumors in dogs. *J. Vet. Intern. Med.* **2008**, *22*, 985–988. [CrossRef] [PubMed]

45. Irie, M.; Takeuchi, Y.; Ohtake, Y.; Suzuki, H.; Nagata, N.; Miyoshi, T.; Kagawa, Y.; Yamagami, T. Imatinib mesylate treatment in a dog with gastrointestinal stromal tumors with a c-kit mutation. *J. Vet. Med. Sci.* **2015**, *77*, 1535–1539. [CrossRef] [PubMed]

46. Kobayashi, M.; Kuroki, S.; Ito, K.; Yasuda, A.; Sawada, H.; Ono, K.; Washizu, T.; Bonkobara, M. Imatinib-associated tumour response in a dog with a non-resectable gastrointestinal stromal tumour harbouring a c-kit exon 11 deletion mutation. *Vet. J.* **2013**, *198*, 271–274. [CrossRef] [PubMed]

47. Bonkobara, M. Dysregulation of tyrosine kinases and use of imatinib in small animal practice. *Vet. J.* **2015**, *205*, 180–188. [CrossRef]

48. Smedley, R.C.; Spangler, W.L.; Esplin, D.G.; Kitchell, B.E.; Bergman, P.J.; Ho, H.Y.; Bergin, I.L.; Kiupel, M. Prognostic markers for canine melanocytic neoplasms: A comparative review of the literature and goals for future investigation. *Vet Pathol.* **2011**, *48*, 54–72. [CrossRef]

49. Bergin, I.L.; Smedley, R.C.; Esplin, D.G.; Spangler, W.L.; Kiupel, M. Prognostic evaluation of Ki67 threshold value in canine oral melanoma. *Vet. Pathol.* **2011**, *48*, 41–53. [CrossRef]

50. Ensembl. Available online: https://www.ensembl.org/index.html (accessed on 9 December 2020).

51. Primer3web. Available online: https://primer3.ut.ee/ (accessed on 9 December 2020).

52. Multiple Sequence Alignment by CLUSTALW. Available online: https://www.genome.jp/tools-bin/clustalw (accessed on 9 December 2020).

53. Fonseca-Alves, C.E.; Kobayashi, P.E.; Palmieri, C.; Laufer-Amorim, R. Investigation of c-KIT and Ki67 expression in normal, preoplastic and neoplastic canine prostate. *Bmc Vet. Res.* **2017**, *13*, 1–9. [CrossRef]

54. Teixeira, T.F.; Da Silva, T.C.; Cogliati, B.; Nagamine, M.K.; Dagli, M.L.Z. Retrospective study of melanocytic neoplasms in dogs and cats. *Braz. J. Vet. Pathol.* **2010**, *3*, 100–104.

55. Eckhart, L.; Bach, J.; Ban, J.; Tschachler, E. Melanin binds reversibly to thermostable DNA polymerase and inhibits its activity. *Biochem. Biophys. Res. Commun.* **2000**, *271*, 726–730. [CrossRef] [PubMed]

56. Lassam, N.; Bickford, S. Loss of c-kit expression in cultured melanoma cells. *Oncogene* **1992**, *7*, 51–56. [PubMed]

57. Montone, K.T.; van Belle, P.; Elenitsas, R.; Elder, D.E. Proto-oncogene c-kit expression in malignant melanoma: Protein loss without tumor progression. *Mod. Pathol.* **1997**, *10*, 939–944.

58. Prouteau, A.; Chocteau, F.; de Brito, C.; Cadieu, E.; Primot, A.; Botherel, N.; Degorce, F.; Cornevin, L.; Lagadic, M.A.; Cabillic, F.; et al. Prognostic value of somatic focal amplifications on chromosome 30 in canine oral melanoma. *Vet. Comp. Oncol.* **2020**, *18*, 214–223. [CrossRef]

Publisher's Note: MDPI stays neutral with regard to jurisdictional claims in published maps and institutional affiliations.

Article

SDHB and SDHA Immunohistochemistry in Canine Pheochromocytomas

Firas M. Abed [1], Melissa A. Brown [2], Omar A. Al-Mahmood [3] and Michael J. Dark [1,2,4,*]

[1] Department of Comparative, Diagnostic and Population Medicine, College of Veterinary Medicine, University of Florida, Gainesville, FL 32610, USA; fabed@ufl.edu

[2] Veterinary Diagnostic Laboratories, College of Veterinary Medicine, University of Florida, Gainesville, FL 32610, USA; melissabrown@ufl.edu

[3] Department of Food, Nutrition and Packaging Science, College of Agriculture, Forestry and Life Sciences, Clemson University, Clemson, SC 29634, USA; oabdull@g.clemson.edu

[4] Emerging Pathogens Institute, University of Florida, Gainesville, FL 32610, USA

* Correspondence: darkmich@ufl.edu; Tel.: +1-(352)-294-4138

Received: 27 August 2020; Accepted: 15 September 2020; Published: 17 September 2020

Simple Summary: Pheochromocytomas are adrenal tumors that occur in both dogs and people. One of the more common gene families involved in the development of this tumor in people is succinate dehydrogenase (SDH). In people, immunohistochemistry can be used with biopsy samples to predict gene pathways that may be involved in the development of the tumor. This is faster and cheaper than performing extensive sequencing to determine if genes are involved. We tested 35 dog tumors to determine how likely SDH mutations were. While our data suggest significant numbers of SDH mutations, these mutations do not appear to be associated with tumor aggression.

Abstract: Pheochromocytomas (PCs) are tumors arising from the chromaffin cells of the adrenal glands and are the most common tumors of the adrenal medulla in animals. In people, these are highly correlated to inherited gene mutations in the succinate dehydrogenase (SDH) pathway; however, to date, little work has been done on the genetic basis of these tumors in animals. In humans, immunohistochemistry has proven valuable as a screening technique for SDH mutations. Human PCs that lack succinate dehydrogenase B (SDHB) immunoreactivity have a high rate of mutation in the SDH family of genes, while human PCs lacking succinate dehydrogenase A (SDHA) immunoreactivity have mutations in the SDHA gene. To determine if these results are similar for dogs, we performed SDHA and SDHB immunohistochemistry on 35 canine formalin-fixed, paraffin-embedded (FFPE) PCs. Interestingly, there was a loss of immunoreactivity for both SDHA and SDHB in four samples (11%), suggesting a mutation in SDHx including SDHA. An additional 25 (71%) lacked immunoreactivity for SDHB, while retaining SDHA immunoreactivity. These data suggest that 29 out of the 35 (82%) may have an SDH family mutation other than SDHA. Further work is needed to determine if canine SDH immunohistochemistry on PCs correlates to genetic mutations that are similar to human PCs.

Keywords: dog; pheochromocytoma; SDH; Immunohistochemistry

---

1. Introduction

Pheochromocytomas (PCs) are catecholamine-secreting neuroendocrine tumors arising from the chromaffin cells of the neural crest [1–3]. PCs are more often seen in dogs and cattle [4] and can be unilateral or bilateral and functional or nonfunctional. While canine pheochromocytomas are usually benign, they can invade adjacent tissues and may be malignant, with metastasis to distant tissues [4]. The behavior of pheochromocytomas is difficult to predict based on histologic findings [5,6].

Immunohistochemistry (IHC) data has found that canine and human PCs are highly similar, as neoplastic cells in both share the expression of numerous antigens, including S100, synaptophysin (SYN), chromogranin A (CGA), and substance P (SP) [7]. Known genetic mutations are involved in the pathogenesis of approximately 60% of human PCs. Frequently, these are associated with mutations in the succinate dehydrogenase (SDH) family of genes, with mutations in succinate dehydrogenase subunit B (SDHB) associated with a high likelihood of metastasis/malignancy [5]. The sequencing of multiple SDH genes in every pheochromocytoma is economically infeasible in veterinary medicine, making the determination of the genetic basis of PC in dogs uncertain.

Studies in human PCs have found that immunohistochemistry (IHC) is highly correlated with the SDH mutation status. For instance, all samples with mutations in SDH family genes lacked SDHB immunoreactivity [8]. Loss of SDHB protein expression has therefore been used for prognostication in human medicine; in one study, the relationship between the SDH genetic background and SDHB immunohistochemistry sensitivity and specificity in human pheochromocytomas was 94.23% [9].

In veterinary medicine, while a link has been established between brachephalic dogs and pheochromocytoma [2], there has not been a definitive familial inheritance pattern like that described in people. Humans and dogs have a similar structure of SDH family genes [10]. Some mutations have been found in SDH family genes, which may indicate that mutations in these genes may initiate oncogenesis in a similar way to people [2,11]. While IHC has previously been used in canine pheochromocytomas to verify the neuroendocrine origin [4,7], only one study to date has examined the SDH family status in canine pheochromocytomas [11]. IHC would be significantly more cost-effective and practical than sequencing for mutation detection. In addition, if IHC is associated with patterns in clinical parameters, such as recurrence, invasion, or metastasis, this could be a valuable adjunct to histopathology.

In this study, we examined the expression of SDHA and SDHB in canine pheochromocytomas and compared these with data on patient age, tumor size, and invasion, in order to determine the utility of SDH IHC in canine pheochromocytoma diagnostics.

2. Materials and Methods

2.1. Approval

This study was approved by the University of Florida Institutional Animal Use and Care Committee (study #201710050).

2.2. Samples

A total of 35 pheochromocytomas plus 40 control tissues (20 sections from the heart and 20 sections from normal adrenal glands) were obtained from paraffin-embedded tissue blocks in the College of Veterinary Medicine Anatomic Pathology, University of Florida tissue archive. Representative sections of each tumor were evaluated by a board-certified veterinary pathologist (MJD) to confirm the diagnosis. Clinical records from all samples from patients of the UF Veterinary Medical Center were analyzed to determine mass size and invasion.

2.3. Immunohistochemistry

Sections from each block were cut at 4 μm and stained using commercially available antibodies (SDHA Mouse Monoclonal Antibody [2E3GC12FB2AE2, ThermoFisher Scientific, catalog #459200] at 1:100 dilution and SDHB Rabbit Polyclonal Antibody [ThermoFisher Scientific, catalog #PA5-23079] at 1:75 dilution). Staining was performed on a Leica Bond immunostainer per the manufacturer's directions. Sections from 20 canine adrenal glands and 20 cardiac samples were used to verify the appropriate immunoreactivity of the antibody with canine antigens.

Pheochromocytoma samples from dogs were listed as positive when there was a granular intracytoplasmic immunoreactivity with a similar intensity as the internal positive controls (endothelial cells, sustentacular cells, and/or lymphocytes). Negative samples lacked immunoreactivity in the cells of the mass that showed immunoreactivity in internal positive controls [9].

2.4. Statistical Analysis

Descriptive statistics were performed using JMP Pro 12 software (SAS Institute Inc., Cary, NC, USA) 21. Descriptive statistics that were calculated included averages and percentiles. The frequencies of positive SDHA and SDHB immunoreactivity between different variables (invasion, sex, breed, age, and animal weight) were evaluated and compared using Chi-square tests. For all statistical analyses, $p > 0.05$ was considered significant.

3. Results

All 40 controls had an intracytoplasmic immunoreactivity for SDHA and SDHB similar to that found in human tissues (Figure 1A,B).

In the 35 sections examined, there was a lack of both SDHA and SDHB immunoreactivity in four samples (11.4%) (Figures 2 and 3); 25 samples (71.4%) lacked SDHB immunoreactivity but had SDHA immunoreactivity. The lack of SDHB immunoreactivity correlated with both age (with younger animals predisposed ($\leq$10 years old: 62.07%, >10 years old: 37.93%; $p < 0.0421$) and sex (male: 75.86%, female: 24.14%; $p < 0.05$).

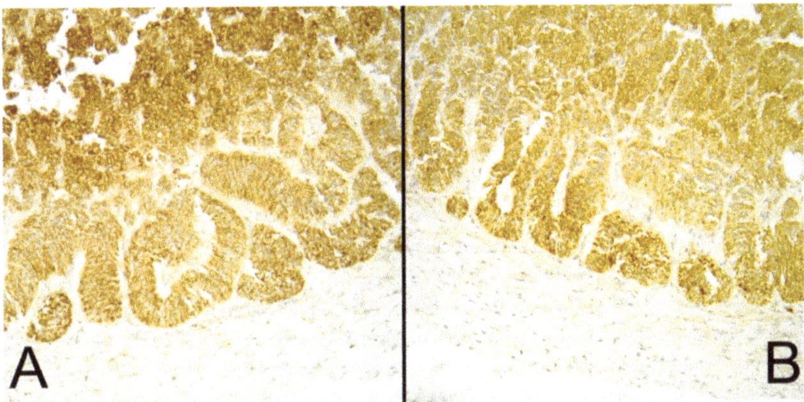

Figure 1. (**A**) Succinate dehydrogenase A (SDHA) and (**B**) Succinate dehydrogenase B (SDHB) immunoreactivity in normal canine adrenal glands as controls.

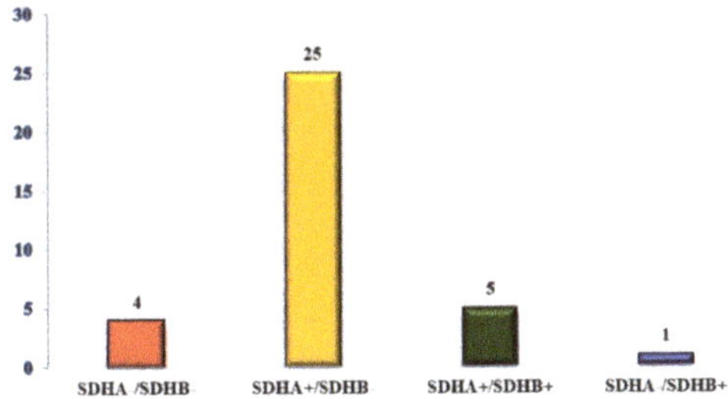

Figure 2. Succinate dehydrogenase A (SDHA)/ Succinate dehydrogenase B (SDHB) immunoreactivity for all samples.

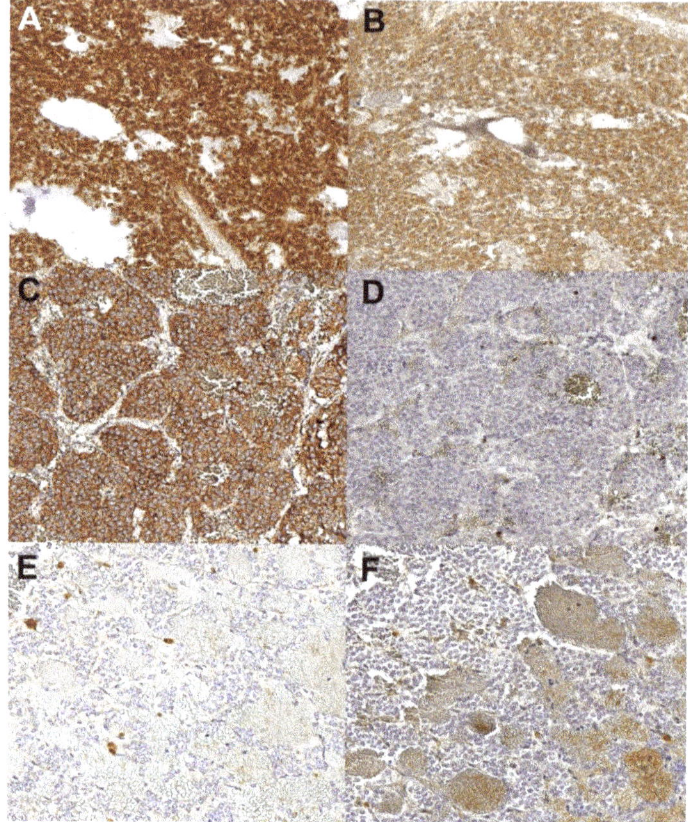

Figure 3. SDHA and SDHB immunoreactivity. (**A,C,E**)—SDHA immunohistochemistry. (**B,D,F**)—SDHB immunoreactivity. (**A**) and (**B**) represent a case with both SDHA and SDHB immunoreactivity; (**C**) and (**D**) have SDHA but lack SDHB immunoreactivity; (**E**) and (**F**) represent a case lacking both SDHA and SDHB immunoreactivity. SDHA—Succinate dehydrogenase A, SDHB—Succinate dehydrogenase B, Original objective 40×.

Out of the 34 cases with associated signalment and clinical information, 12 were from females and 24 were from males (Table 1). The mean patient age was 10.4 years (standard deviation: 2.5 years). Out of 23 samples that had a clinical evidence of invasion, 19 (82.6%) lacked SDHB immunoreactivity. Of these, 14 samples (73.6%) had a vascular invasion (caudal vena cava, phrenicoabdominal vein, and/or renal vein), while two out of these 19 (10.53%) metastasized to the renal vein or regional lymph node. Out of the 12 samples with no reported invasion, 10 (34.84%) lacked immunoreactivity to SDHB. However, we lacked an intraoperative surgical report for four of these samples, which means that these may have an invasion that was not stated in the sample submission form. Two samples out of these 12 described adhesion to the omentum, splenic vessels, apex of the left pancreatic limb, and/or the left kidney. Out of the five samples that had both SDHA and SDHB immunoreactivity, three (60%) also had an invasion. There was no significant correlation between invasions and the immunohistochemical status, age, sex, breed, or animal weight.

Table 1. Immunohistochemical findings and clinical information for all cases.

Case	SDHA	SDHB	Invasion	Age	Sex	Breed
1	–	–	Caudal vena cava	10	M	Mixed Breed
2	–	–	None	8	M	Carolina Dog
3	–	–	Caudal vena cava	14	M	Mixed Breed
4	–	–	Caudal vena cava	8	M	Rhodesian Ridgeback
5	+	+	None	11	M	Terrier, Yorkshire
6	+	+	Caudal vena cava	13	F	Terrier, Scottish
7	+	+	Caudal vena cava	14	F	Mixed Breed
8	+	+	Phrenicoabdominal vein	11	F	Cocker Spaniel
9	+	+	None	8	F	Bassett Hound
10	+	–	Caudal vena cava, left external iliac	9	M	Mixed Breed
11	+	–	Caudal vena cava	9	M	Australian Shepherd
12	+	–	None	11	F	Schnauzer, Miniature
13	+	–	Caudal vena cava, left renal vein	13	F	Mixed Breed
14	+	–	Caudal vena cava	8	F	Rhodesian Ridgeback
15	+	–	Adhered to the dorsal body wall	12	M	Mixed Breed
16	+	–	None	7	F	Spaniel, Boykin
17	+	–	Phrenicoabdominal vein	12	M	Retriever, Golden
18	+	–	Adhered to the left pancreatic limb and the left kidney	16	M	Mixed Breed
19	+	–	Caudal vena cava, phrenicoabdominal vein	9	F	Miniature Schnauzer
20	+	–	Caudal vena cava, phrenicoabdominal vein	12	M	Jack Russell Terrier
21	+	–	None	9	M	Australian Blue Heeler
22	+	–	Caudal vena cava	8.4	M	Shetland Sheepdog
23	+	–	None	8.5	M	Rhodesian Ridgeback
24	+	–	Caudal vena cava, phrenicoabdominal vein	8.1	F	Doberman Pinscher
25	+	–	Phrenicoabdominal vein	11.8	M	Beagle
26	+	–	Caudal vena cava	9.7	M	Fox terrier
27	+	–	None	13	M	Dachshund
28	+	–	Mesenteric lymph node metastasis	11	M	Terrier, Boston
29	–	+	Renal vein	15	F	Retriever, Golden
30	+	–	None	12	M	Dachshund, Miniature
31	+	–	None	10	M	Rottweiler
32	+	–	Phrenicoabdominal vein	7	M	Schnauzer, Standard
33	+	–	No clinical information provided			
34	+	–	Caudal vena cava	5	F	Mixed Breed
35	+	–	Caudal vena cava	11	M	Mixed Breed

SDHA—Succinate dehydrogenase A, SDHB—Succinate dehydrogenase B, M—Male, F—Female.

4. Discussion

Human studies have found that 75% of samples with SDHA mutations lacked immunoreactivity for SDHA and SDHB [9] and that 90% of samples with SDHB, C, D, or SDHAF2 mutations had immunoreactivity for SDHA and lacked immunoreactivity for SDHB [9,12,13]. When this is applied to the samples in our study, these patterns suggest that 29 out of the 35 (82.8%) samples have a mutation in at least one of the SDH family genes (Figure 2).

The predominance of samples with an invasion lacking SDHB immunoreactivity agrees with several human studies [5,12,14,15], which show that a lack of SDHB immunoreactivity is associated with more aggressive tumor behavior.

Only one sample lacked SDHA immunoreactivity but maintained SDHB immunoreactivity (Figure 1); this sample showed signs of invasion and metastasis to the renal vein. However, as inactivation of SDHA does not lead to tumorigenesis [16], this may represent a somatic hypermethylation of the promoter region [17], leading to an accumulation of succinate and later an inhibition of demethylase enzymes. This can lead to promoter hypermethylation and tumor suppressor gene inactivation [18]; in humans, this is known as Leigh syndrome.

The lack of significant association between the immunohistochemical findings and tissue invasion may be due to limited numbers of cases with both SDHA and SDHB immunoreactivity.

Ultimately, while these data are encouraging, sequencing is needed to fully determine the role of SDH family genes in pheochromocytoma in canine pheochromocytomas. These findings do suggest that IHC has a similar utility in narrowing candidate mutations in pheochromocytomas in dogs; given, in particular, the relatively higher cost of sequencing, restricting the set of possible candidate genes using immunohistochemistry would be helpful in future sequencing studies. However, determining the true utility of SDH immunohistochemistry would require sequencing information and additional case information; if immunohistochemistry is associated with specific mutations and/or mutations in specific genes are closely associated with clinical behavior, immunohistochemistry would become an important tool in case management.

5. Conclusions

Canine pheochromocytomas have similar immunohistochemical characteristics to those previously reported in human tumors, with approximately 82% showing immunohistochemical evidence of an SDH family mutation. However, based on a small number of cases, there does not appear to be a correlation with invasion and SDH family gene mutation status. Further sequencing work is needed to verify the IHC results and to determine the genetic basis of canine pheochromocytoma; a larger pool of cases is also needed to confirm a lack of association between invasion and SDH family mutation status.

Author Contributions: Conceptualization, F.M.A. and M.J.D.; methodology, F.M.A., M.A.B., O.A.A.-M., M.J.D.; formal analysis, F.M.A., O.A.A.-M., M.J.D.; investigation, F.M.A., M.A.B., O.A.A.-M., M.J.D.; writing—original draft preparation, F.M.A., M.J.D.; writing—review and editing, F.M.A., M.A.B., O.A.A.-M., M.J.D.; visualization, M.J.D.; supervision, M.J.D. All authors have read and agreed to the published version of the manuscript.

Funding: This research received no external funding.

Conflicts of Interest: The authors declare no conflict of interest.

References

1. Comino-Mendez, I.; de Cubas, A.A.; Bernal, C.; Alvarez-Escola, C.; Sanchez-Malo, C.; Ramirez-Tortosa, C.L.; Pedrinaci, S.; Rapizzi, E.; Ercolino, T.; Bernini, G.; et al. Tumoral EPAS1 (HIF2A) mutations explain sporadic pheochromocytoma and paraganglioma in the absence of erythrocytosis. *Hum. Mol. Genet.* **2013**, *22*, 2169–2176. [CrossRef]
2. Holt, D.E.; Henthorn, P.; Howell, V.M.; Robinson, B.G.; Benn, D.E. Succinate dehydrogenase subunit D and succinate dehydrogenase subunit B mutation analysis in canine phaeochromocytoma and paraganglioma. *J. Comp. Pathol* **2014**, *151*, 25–34. [CrossRef] [PubMed]
3. Wilzen, A.; Rehammar, A.; Muth, A.; Nilsson, O.; Tesan Tomic, T.; Wangberg, B.; Kristiansson, E.; Abel, F. Malignant pheochromocytomas/paragangliomas harbor mutations in transport and cell adhesion genes. *Int. J. Cancer* **2016**, *138*, 2201–2211. [CrossRef] [PubMed]
4. Barthez, P.Y.; Marks, S.L.; Woo, J.; Feldman, E.C.; Matteucci, M. Pheochromocytoma in dogs: 61 cases (19841–995). *J. Vet. Intern. Med.* **1997**, *11*, 272–278. [CrossRef] [PubMed]
5. Blank, A.; Schmitt, A.M.; Korpershoek, E.; van Nederveen, F.; Rudolph, T.; Weber, N.; Strebel, R.T.; de Krijger, R.; Komminoth, P.; Perren, A. SDHB loss predicts malignancy in pheochromocytomas/sympathethic paragangliomas, but not through hypoxia signalling. *Endocr. Relat. Cancer* **2010**, *17*, 919–928. [CrossRef] [PubMed]
6. Thompson, L.D. Pheochromocytoma of the Adrenal gland Scaled Score (PASS) to separate benign from malignant neoplasms: A clinicopathologic and immunophenotypic study of 100 cases. *Am. J. Surg. Pathol.* **2002**, *26*, 551–566. [CrossRef] [PubMed]
7. Sako, T.; Kitamura, N.; Kagawa, Y.; Hirayama, K.; Morita, M.; Kurosawa, T.; Yoshino, T.; Taniyama, H. Immunohistochemical evaluation of a malignant phecochromocytoma in a wolfdog. *Vet. Pathol.* **2001**, *38*, 447–450. [CrossRef] [PubMed]
8. Menara, M.; Oudijk, L.; Badoual, C.; Bertherat, J.; Lepoutre-Lussey, C.; Amar, L.; Iturrioz, X.; Sibony, M.; Zinzindohoue, F.; de Krijger, R.; et al. SDHD immunohistochemistry: A new tool to validate SDHx mutations in pheochromocytoma/paraganglioma. *J. Clin. Endocrinol. Metab.* **2015**, *100*, E287–E291. [CrossRef] [PubMed]
9. Papathomas, T.G.; Oudijk, L.; Persu, A.; Gill, A.J.; van Nederveen, F.; Tischler, A.S.; Tissier, F.; Volante, M.; Matias-Guiu, X.; Smid, M.; et al. SDHB/SDHA immunohistochemistry in pheochromocytomas and paragangliomas: A multicenter interobserver variation analysis using virtual microscopy: A Multinational Study of the European Network for the Study of Adrenal Tumors (ENS@T). *Mod. Pathol.* **2015**, *28*, 807–821. [CrossRef] [PubMed]

10. Rustin, P.; Munnich, A.; Rotig, A. Succinate dehydrogenase and human diseases: New insights into a well-known enzyme. *Eur. J. Hum. Genet.* **2002**, *10*, 289–291. [CrossRef] [PubMed]

11. Korpershoek, E.; Dieduksman, D.; Grinwis, G.C.M.; Day, M.J.; Reusch, C.E.; Hilbe, M.; Fracassi, F.; Krol, N.M.G.; Uitterlinden, A.G.; de Klein, A.; et al. Molecular Alterations in Dog Pheochromocytomas and Paragangliomas. *Cancers* **2019**, *11*, 607. [CrossRef]

12. Gill, A.J.; Benn, D.E.; Chou, A.; Clarkson, A.; Muljono, A.; Meyer-Rochow, G.Y.; Richardson, A.L.; Sidhu, S.B.; Robinson, B.G.; Clifton-Bligh, R.J. Immunohistochemistry for SDHB triages genetic testing of SDHB, SDHC, and SDHD in paraganglioma-pheochromocytoma syndromes. *Hum. Pathol.* **2010**, *41*, 805–814. [CrossRef] [PubMed]

13. Van Nederveen, F.H.; Gaal, J.; Favier, J.; Korpershoek, E.; Oldenburg, R.A.; de Bruyn, E.M.; Sleddens, H.F.; Derkx, P.; Riviere, J.; Dannenberg, H.; et al. An immunohistochemical procedure to detect patients with paraganglioma and phaeochromocytoma with germline SDHB, SDHC, or SDHD gene mutations: A retrospective and prospective analysis. *Lancet. Oncol.* **2009**, *10*, 764–771. [CrossRef]

14. Casey, R.T.; Ascher, D.B.; Rattenberry, E.; Izatt, L.; Andrews, K.A.; Simpson, H.L.; Challis, B.; Park, S.M.; Bulusu, V.R.; Lalloo, F.; et al. SDHA related tumorigenesis: A new case series and literature review for variant interpretation and pathogenicity. *Mol. Genet. Genomic Med.* **2017**, *5*, 237–250. [CrossRef] [PubMed]

15. Pinato, D.J.; Ramachandran, R.; Toussi, S.T.; Vergine, M.; Ngo, N.; Sharma, R.; Lloyd, T.; Meeran, K.; Palazzo, F.; Martin, N.; et al. Immunohistochemical markers of the hypoxic response can identify malignancy in phaeochromocytomas and paragangliomas and optimize the detection of tumours with VHL germline mutations. *Br. J. Cancer* **2013**, *108*, 429–437. [CrossRef] [PubMed]

16. Eng, C.; Kiuru, M.; Fernandez, M.J.; Aaltonen, L.A. A role for mitochondrial enzymes in inherited neoplasia and beyond. *Nat. Rev. Cancer* **2003**, *3*, 193–202. [CrossRef] [PubMed]

17. Haller, F.; Moskalev, E.A.; Faucz, F.R.; Barthelmess, S.; Wiemann, S.; Bieg, M.; Assie, G.; Bertherat, J.; Schaefer, I.M.; Otto, C.; et al. Aberrant DNA hypermethylation of SDHC: A novel mechanism of tumor development in Carney triad. *Endocr. Relat. Cancer* **2014**, *21*, 567–577. [CrossRef] [PubMed]

18. Letouze, E.; Martinelli, C.; Loriot, C.; Burnichon, N.; Abermil, N.; Ottolenghi, C.; Janin, M.; Menara, M.; Nguyen, A.T.; Benit, P.; et al. SDH mutations establish a hypermethylator phenotype in paraganglioma. *Cancer Cell* **2013**, *23*, 739–752. [CrossRef] [PubMed]

Article

Canine Gastric Carcinomas: A Histopathological and Immunohistochemical Study and Similarities with the Human Counterpart

Alexandros Hardas [1,*], Alejandro Suárez-Bonnet [1], Sam Beck [2], William E. Becker [1], Gustavo A. Ramírez [3] and Simon L. Priestnall [1]

[1] Department of Pathobiology & Population Sciences, The Royal Veterinary College, North Mymms, Hatfield, Hertfordshire AL9 7TA, UK; asuarezbonnet@rvc.ac.uk (A.S.-B.); wbecker6@rvc.ac.uk (W.E.B.); spriestnall@rvc.ac.uk (S.L.P.)
[2] VPG Histology, Horfield, Bristol BS7 0BJ, UK; sam.beck@synlab.co.uk
[3] Department of Animal Science, School of Agriculture, Food Science and Veterinary Medicine (ETSEA), University of Lleida, 25198 Lleida, Spain; gustavo.ramirez@udl.cat
* Correspondence: achardas@rvc.ac.uk

Simple Summary: Gastric carcinoma (GC) continues to be one of the leading causes of death in humans and is the most common neoplasm in the stomachs of dogs. In both species, previous studies have demonstrated that the disease is heterogeneous, with genetic and environmental factors playing a quintessential role in disease pathogenesis. Compared to humans, the incidence of gastric carcinoma in dogs is low although, in a small number of breeds, a higher incidence has been reported. In dogs, the etiology and molecular pathways involved remain largely unknown. This retrospective study reviews current signalment data, evaluates the inflammatory component and association with *Helicobacter* spp. presence in various canine gastric carcinoma histological subtypes, and investigates potential molecular pathways involved in one of the largest study cohorts to date. The benefit of such a comparative study is to highlight the parallel histological features and molecular pathways between dogs and humans.

Citation: Hardas, A.; Suárez-Bonnet, A.; Beck, S.; Becker, W.E.; Ramírez, G.A.; Priestnall, S.L. Canine Gastric Carcinomas: A Histopathological and Immunohistochemical Study and Similarities with the Human Counterpart. *Animals* **2021**, *11*, 1409. https://doi.org/10.3390/ani11051409

Academic Editor: Adelina Gama

Received: 10 March 2021
Accepted: 10 May 2021
Published: 14 May 2021

Publisher's Note: MDPI stays neutral with regard to jurisdictional claims in published maps and institutional affiliations.

Abstract: Canine gastric carcinoma (CGC) affects both sexes in relatively equal proportions, with a mean age of nine years, and the highest frequency in Staffordshire bull terriers. The most common histological subtype in 149 CGC cases was the undifferentiated carcinoma. CGCs were associated with increased chronic inflammation parameters and a greater chronic inflammatory score when *Helicobacter* spp. were present. Understanding the molecular pathways of gastric carcinoma is challenging. All markers showed variable expression for each subtype. Expression of the cell cycle regulator 14-3-3σ was positive in undifferentiated, tubular and papillary carcinomas. This demonstrates that 14-3-3σ could serve as an immunohistochemical marker in routine diagnosis and that mucinous, papillary and signet-ring cell (SRC) carcinomas follow a 14-3-3σ independent pathway. p16, another cell cycle regulator, showed increased expression in mucinous and SRC carcinomas. Expression of the adhesion molecules E-cadherin and CD44 appear context-dependent, with switching within tumor emboli potentially playing an important role in tumor cell survival, during invasion and metastasis. Within neoplastic emboli, acinar structures lacked expression of all markers, suggesting an independent molecular pathway that requires further investigation. These findings demonstrate similarities and differences between dogs and humans, albeit further clinicopathological data and molecular analysis are required.

Keywords: canine; stomach; gastric carcinoma; *Helicobacter* spp.; p16; 14-3-3σ; E-cadherin and CD44

1. Introduction

Gastric carcinoma is a rare form of cancer in domestic animals, and in dogs accounts for <1% of all reported neoplasms [1], with adenocarcinoma the most frequent (50–90%) [1]. The median age of gastric carcinoma development is 10 years, with rough collies, Staffordshire bull terriers, chow-chows, Belgian shepherds, Norwegian Lundehunds, Cairn terriers and West Highland white terriers being the breeds most likely to be affected [2].

Compared to gastric carcinoma in humans, the incidence of canine gastric carcinoma (CGC) is relatively low; however, in recent years the disease has been more frequently diagnosed [3,4]. Considering the proposed causal effect of diet on gastric neoplasia in humans [5], this increased frequency in dogs over the last 30 years could be similarly attributed. Breed predisposition has also contributed [6], with those breeds at increased risk becoming more popular. In addition, increased longevity and advances in veterinary diagnostic techniques, such as gastroscopy, may have also contributed to increased diagnosis.

Clinical signs of CGC include vomiting that may progress to hematemesis, melaena, anemia, lethargy, ptyalism, polydipsia, abdominal distension, and abdominal discomfort. Prognosis is generally poor, with a median survival time of 35 days, and confirmed metastasis in about 70–90% of cases at the time of diagnosis or death [7]. Common sites of metastasis include the gastric lymph nodes, omentum, liver, duodenum, pancreas, spleen, esophagus, adrenal glands and lungs [2].

Classification of gastric carcinomas in dogs follows the World Health Organization (WHO) [8] scheme, adapted from humans, which is based on the predominant histological features and the main patterns of cells within the neoplasm: papillary, tubular, mucinous, signet-ring cell (SRC) and undifferentiated types [2]. An alternative scheme, again adapted from human medicine, the Lauren classification, divides tumors into intestinal—cohesive masses and tubular structures; diffuse—individual or scattered nests of neoplastic cells; and mixed—incorporating features of both intestinal and diffuse types [9]. Both schemes have been applied in previous studies [1,3].

The association between chronic inflammation, caused by a variety of factors (bacterial, viral, and parasitic infections, chemical irritants, and nondigestible particles), and carcinogenesis is now well established in humans and animals [10]. The risk of carcinogenesis is higher, the longer the inflammation persists [11]. In humans, it has been shown that several risk factors, such as *Helicobacter pylori* infection, diet, and smoking, are involved in the precancerous cascade of events that lead to gastric adenocarcinomas [12].

The pathogenesis of CGC remains elusive, albeit a high prevalence in certain breeds (e.g., Staffordshire bull terrier, Norwegian Lundehund and Belgian shepherd dog) suggests an underlying genetic etiology [6]. A clear role for *Helicobacter* spp., similar to *H. pylori* in humans, has not been reported in domestic species to date [13]. Whether other *Helicobacter* spp. are involved with carcinogenesis is unclear. *H. pylori* has occasionally been recognized in the canine stomach [2,12], however, the predominant species in dogs are *H. felis*, *H. bizzozeronii* and *H. heilmannii* [14,15]. A clear association between gastric inflammation and *Helicobacter* spp. presence has not been made in previous studies, and in addition, an association with gastric carcinoma has also not been investigated [16].

The role of cell cycle regulators and cell adhesion molecules in cancer is complex and paradoxical, varying by cell type and stage of tumorigenesis [17]. In this study, we aimed to examine the involvement of important cell cycle regulators and cell adhesion molecules, previously studied in human gastric carcinoma, in dogs.

E-cadherin is a calcium-dependent cell–cell adhesion molecule that preserves epithelial integrity and can act both as a tumor-suppressor and as an oncoprotein [18,19]. A recent large-scale study separating subtypes according to their growth pattern (polypoid or non-polypoid, i.e., signet cell, mucinous and undifferentiated carcinoma) showed that there is a complete loss of E-cadherin in non-polypoid and undifferentiated carcinomas, and reduced expression in polypoid, with no evidence of malignant alteration or invasion in canine gastrointestinal tumors [20].

CD44 is a cell surface receptor for hyaluronic acid and binds to collagen, fibronectin and chondroitin sulfate [21]. Its role in tumorigenesis and metastasis is thought to be through signaling pathways that regulate cell adhesion, migration, proliferation, differentiation and survival [22,23]. Histopathological studies of human gastric carcinoma have associated high CD44 expression with tumor invasion, lymph node metastasis and patient survival [24–27], although the expression of CD44 in CGC has so far not been investigated.

p16 protein inhibits cyclinD-CDK4/6, and previous studies of human gastric carcinoma have shown that loss of p16 expression has been associated with increased measures of malignancy and poor clinical outcome [28]. One previous study showed loss of expression of p16 in seventeen cases of CGC [29].

14-3-3σ is the focus of much research in human medicine, including gastric carcinomas [30,31]. 14-3-3σ protein regulates the G1/S and G2/M cell cycle checkpoints through sequestration of CDK4, CDK2, and CDK1, and thus prevents mitosis and allows DNA repair. It may act as a tumor suppressor [32] or it may serve as an oncoprotein. Previously, in veterinary species, 14-3-3σ has been reported as an oncoprotein in canine mammary and urinary bladder carcinomas [33,34]. The association and implication of 14-3-3σ with CGC will be examined later in this study.

The aims of this study were to provide an update on the signalment data and histopathological classification of a large case series of CGC, and to further investigate the potential association of chronic inflammation and the presence of *Helicobacter* spp. with cancer. Furthermore, using a subset of cases, the expression patterns of four proteins (E-cadherin, p16, 14-3-3σ and CD44) were studied to determine their potential involvement in CGC development.

2. Materials and Methods

2.1. Case Selection

The surgical biopsy databases at VPG Histology, United Kingdom, and the SIDAVE-University of Lleida, Spain were searched for cases of CGC from 2009 to 2019. All cases included relevant medical records (signalment, clinical history, gross description, microscopic description and original diagnosis), and tissues were received at the Royal Veterinary College (RVC) as formalin-fixed paraffin-embedded wax blocks. Individually identifiable owner information was redacted by both supplying institutions. Cases with insufficient tissue or where the diagnosis was not certain on review were excluded.

A control group of canine gastric biopsy samples was established from cases provided by VPG and from the RVC pathology archive. Control group samples were defined based on sample quality, absence of gastric carcinoma, and no previous history of gastric carcinoma. Controls were animals presenting with typical gastrointestinal clinical signs including vomiting, diarrhea and weight loss; however, no tumor was present on histopathological examination.

2.2. Histopathological Evaluation

Histological sections, cut at 4 μm and stained with hematoxylin and eosin, were produced from the provided paraffin blocks and were evaluated microscopically by three veterinary pathologists (A.H., A.S.-B. and S.L.P.). Each case was confirmed as gastric carcinoma and further classified according to the WHO standard into five categories: tubular, papillary, mucinous, SRC, or undifferentiated carcinoma [35,36]. The most frequent histological pattern served as the main classification criterion for each case.

Mucosal inflammation present within histological sections was assessed using modified World Small Animal Veterinary Association standards [8,37,38]. For the purposes of this study, only those parameters consistent with chronic inflammation (intraepithelial lymphocytes, lamina propria lymphocytes and plasma cells, and gastric lympho-follicular hyperplasia) were scored as normal—0, mild—1, moderate—2, and marked—3, following modified WSAVA criteria [37]. A total chronic inflammation score (TCIS, ranging from 0 to 9) was recorded as the sum of the three parameters. Regions of surface ulceration were avoided when assessing chronic inflammatory parameters.

Sections were stained with Warthin–Starry for *Helicobacter* spp. identification. Quantification of *Helicobacter* spp. was performed based on the presence or absence of bacteria in the mucosa using the following system: 0—no helicobacter found in the sample; 1—low numbers of helicobacter found (<15 bacteria per field of highest helicobacter density); and 2—high numbers of helicobacter found (>15 bacteria per field).

2.3. Immunohistochemistry

Immunohistochemical labeling for E-cadherin, CD44, p16 and 14-3-3σ was performed on 4-μm-thick sections mounted on positively charged slides (SuperFrost Plus; Menzel Gläser, Braunschweig, Germany). Antigen retrieval, labeling, and counterstaining were performed on a Bond-Max Autostainer (Leica Biosystems, Newcastle-upon-Tyne, UK) using the Bond Polymer Refine detection system (Leica Biosystems). Primary antibodies and retrieval conditions were as follows: p16 (PA0016, 1:100,Novocastra, Newcastle Upon Tyne, UK); pH 6.0 buffer (ER1, Leica Biosystems) for 20 min, 14-3-3σ (SC-100638, 1:40, Santa Cruz Biotechnology, Heidelberg, Germany); pH 6.0 buffer (ER1) for 20 min, E-cadherin (NCH-38, 1:100, Agilent, Stockport, UK); pH 6.0 buffer (ER1) for 20 min, and CD44 (ab157107, 1:50 Abcam, Cambridge, UK); pH 9.0 buffer (ER2, Leica Biosystems) for 20 min. Internal positive tissue controls for E-cadherin and CD44 were available on each section.

2.4. Evaluation of E-Cadherin, CD44, p16 and 14-3-3σ Immunolabeling

Analysis of immunolabeled samples was performed by 3 veterinary pathologists (A.H., A.S.-B. and S.L.P.), discrepancies were discussed by use of a multi-headed microscope, and a consensus was reached. In all cases, areas of ulceration and/or necrosis were avoided for interpretation.

The distribution and intensity of E-cadherin and CD44 labeling were analyzed for each case using a similar semiquantitative scoring system. The percentages of tumor cells, neoplastic acini and intravascular tumor emboli that expressed CD44 and E-cadherin were assessed. Labeling for CD44 was scored as follows: 0, no labeling; 1, 1–9% positive tumor cells; 2, 10–49% cells; and 3, 50–99% cells. Labeling for E-cadherin was scored as follows: 0, no labeling; 1, 1–9% positive tumor cells; 2, 10–49% cells; 3, 50–79% cells; and 4, 80–100% cells. The intensity of E-cadherin and CD44 labeling (0—negative, 1—mild, 2—moderate, and 3—strong) and whether labeling was membranous and/or cytoplasmic was also recorded [39–41]. A total immunohistochemical score (TIS) [39] for E-cadherin (ranging from 0 to 12) and TIS CD44 (ranging from 0 to 9) was calculated as the product of the distribution and intensity scores. Normal gastric surface epithelium served as a positive internal control for E-cadherin and lymphoid follicles, satellite glial cells and macrophages for CD44.

For p16, neoplastic cells with cytoplasmic and/or nuclear antigen expression were interpreted as immunopositive. An immunohistochemical score was assigned based on the intensity of staining (0—none, 1—weak, 2—moderate, and 3—strong) and percentage of positive tumor cells (0, 0% cells; 1, <25% cells; 2, 25–50% cells; 3, 51–75% cells; 4, >75% cells) [42]. The product of the intensity and distribution scores gave a TIS p16 ranging from 0 to 12. As a positive control, normal skin was used where epithelial cells within the basal layer of the epidermis exhibited weak immunoreactivity.

14-3-3σ labeling was assessed using a previously published semiquantitative scoring system [43]. The scores for the percentage (1, ≤10% cells; 2, 11–50% cells; and 3, >50% cells) and staining intensity (0, negative; 1, weak; 2, moderate; and 3, intense) of positive cells were recorded and a TIS 14-3-3σ (ranging from 0 to 9) was calculated as the product of these two parameters for each of the studied cases. Normal canine urinary bladder, which showed cytoplasmic immunolabeling of the urothelium, was used as a positive control.

2.5. Statistical Analysis

Results were analyzed for any significant relationship between the following: signalment (i.e., age, sex and breed); biopsy type (endoscopic vs. full-thickness biopsies); and histopathological features (tumor classification, chronic inflammation score, helicobacter identification, quantification, and localization). Statistical comparisons were performed using GraphPad Prism version 8.0 for Windows (GraphPad Software, San Diego, CA, USA). Bar graphs were constructed to depict the mean and standard error of the parameter assessed for each group. To test for the significance of a relationship between two categorical variables, a Fisher's test was used to analyze the significance of contingency tables such as endoscopic vs. full-thickness biopsies for finding helicobacter. An unpaired *t*-test was used to compare the mean inflammation scores of different groups. For continuous data, the Student's *t*-test, analysis of variance (ANOVA), or Mann–Whitney U analysis test was used, depending on whether the data were normally distributed and whether two, or more than two groups, were compared. The significance level for all statistical tests was set at $p < 0.05$. The expression of p16, 14-3-3σ, E-cadherin and CD44 were compared with histological subtype and features of malignancy.

3. Results

3.1. Case Details and Signalment

One hundred and eighty-two (182) cases of CGC were identified from the archives of VPG Histology and the University of Lleida and submitted to the Department of Pathobiology and Population Sciences at the Royal Veterinary College. Following the initial histopathological review, 149 were suitable for inclusion in the study based on sample quality, presence of gastric carcinoma, and amount of tumor present in the sample. Cases included 85 (57.0%) endoscopically obtained (mucosa only) samples, and 64 (44.0%) full-thickness biopsies collected by surgical excision. Of the 149 dogs, 69 (46.3%) were female, 75 (50.3%) male and 5 (3.4%) of unreported sex. Neuter status was known in 118 cases, with 88 (74.6%) neutered. Twenty-nine non-tumor control cases were included, 24 (82.8%) endoscopic and 5 (3.4%) full-thickness biopsies. The control group comprised 15 males (51.7%), 13 females (44.8%), and one of unreported sex (3.4%). For the carcinoma group the mean and median ages were 9.1 and 9 years, respectively (range 1–16 years), and for the control group were 7.5 and 7, respectively.

The Staffordshire bull terrier (22/149, 14.8%) was the breed most frequently affected by gastric carcinoma. Of the other non-cross breeds, Labrador retrievers (12, 8.1%), golden retrievers (10, 6.7%), Boxers (9, 6.0%), Border collies (8, 5.4%), rough collies (6, 4.0%), and Belgian shepherd dogs (6, 4.0%) were also commonly affected (Table S1). A similar range of breeds was seen in the control population, with Staffordshire bull terriers, Labrador retrievers, Border collies and Belgian shepherd dogs all represented. No statistical correlation with breed, sex and age was found with regards to each CGC subtype.

3.2. Histopathological Subtypes

Undifferentiated carcinoma was the most frequent carcinoma type overall, comprising 59/149 (39.6%) samples (Table 1). SRC carcinoma made up 47 (31.5%), tubular 32 (21.5%), and mucinous 10 (6.7%). The papillary type was rare, with only a single case (0.7%) (Figure 1) There were no significant associations identified between breed or sex and any of the five classifications of gastric carcinoma.

Table 1. Canine gastric carcinoma subtype by sex.

Gastric Carcinoma Subtype	Male	Female	Not Recorded	Total
Undifferentiated	32	26	1	59 (39.6%)
Signet-ring	22	24	1	47 (31.5%)
Tubular	17	12	3	32(21.5%)
Mucinous	4	6	-	10 (6.7%)
Papillary	0	1	-	1 (0.7%)
Total	75	69	5	149

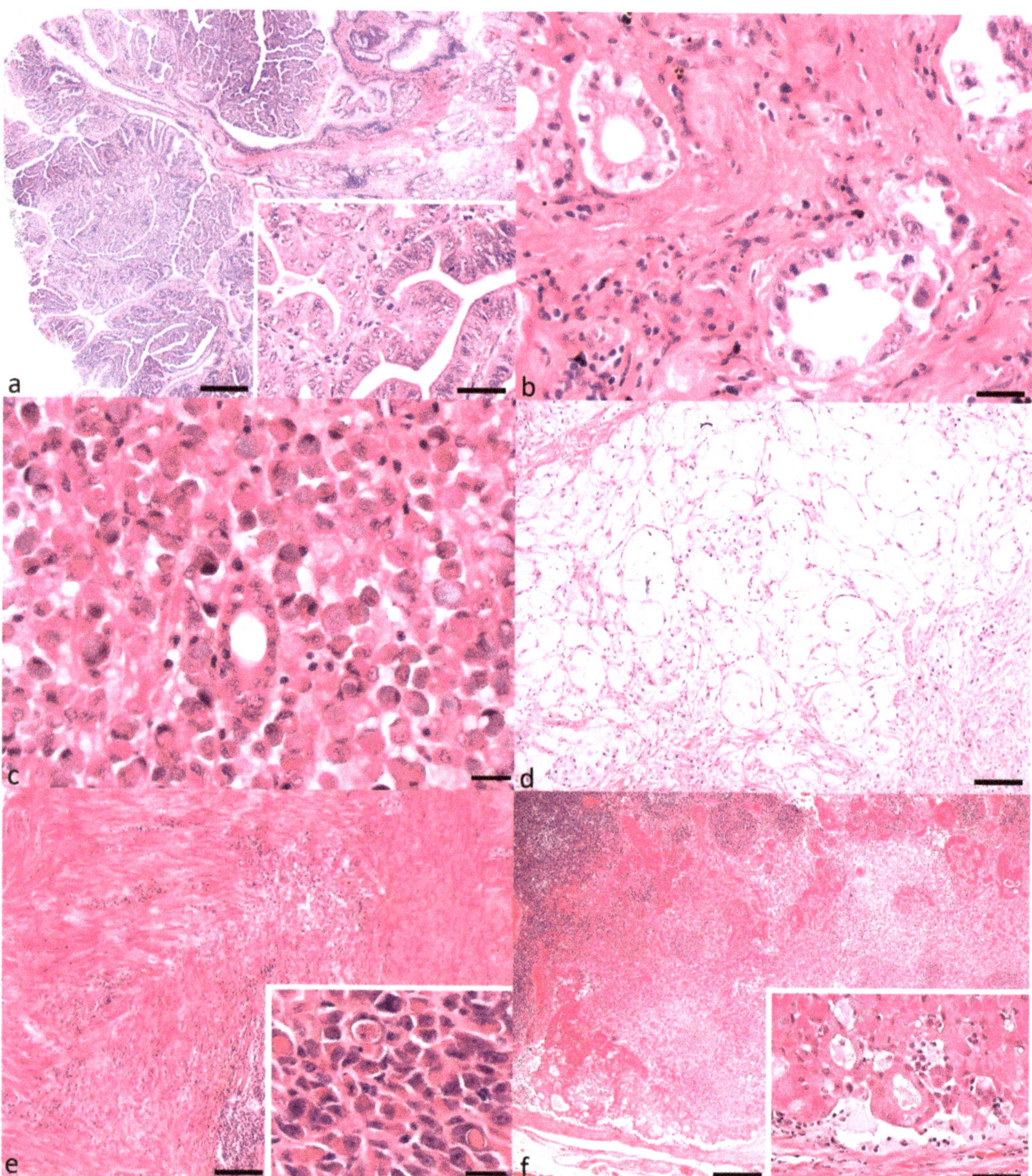

Figure 1. Gastric carcinoma subtypes and local lymph node metastasis. (**a**) Papillary adenocarcinoma with characteristic fibrovascular stalks supporting neoplastic cells that form fingerlike projections (bar = 500 μm). Inset shows increased mitoses (bar = 20 μm). (**b**) Tubular adenocarcinoma with neoplastic tubules lined by pleomorphic cells (bar = 20 μm). (**c**) Signet-ring cell carcinoma with characteristic signet-ring cells diffusely replacing gastric mucosa (bar = 20 μm). (**d**) Mucinous adenocarcinoma with abundant lakes of mucin and occasional small numbers of signet-ring cells (bar = 50 μm). (**e**) Undifferentiated adenocarcinoma with neoplastic cells replacing and dissecting through muscularis layers (bar = 200 μm). Inset showing entotic cell-in-cell (CIC) patterns in an undifferentiated adenocarcinoma (bar = 20 μm). (**f**) Lymph node metastasis of a tubular adenocarcinoma with multifocal intravascular neoplastic emboli (bar = 200 μm). Inset shows a high-power magnification of neoplastic tubule formation within the neoplastic emboli (bar = 20 μm).

Of the 85 endoscopic samples, 36 (42.4%) were SRC carcinoma, 25 (29.4%) were undifferentiated, 19 (22.3%) tubular, and 5 (5.8%) mucinous. Of the 64 full-thickness samples, 34 (53.1%) were undifferentiated carcinoma, 13 (20.3%) tubular, 11 (17.2%) SRC, and 5 (7.7%) mucinous. The single papillary adenocarcinoma was recorded amongst the full-thickness samples (1.5%) (Figure 2). The prevalence of SRC among endoscopic biopsies was significantly higher than among full-thickness biopsies ($p = 0.0002$). Comparisons between other subtypes in endoscopic and full-thickness biopsies did not show any statistical significance.

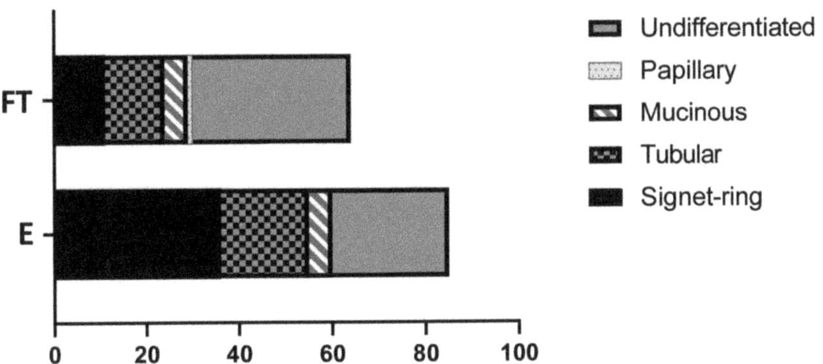

Figure 2. WHO classification of 149 canine gastric carcinomas separated by biopsy technique. Full-thickness surgical (FT), endoscopic (E).

3.3. Inflammation and Presence of Helicobacter spp.

Mean total chronic inflammation score (TCIS) for the control group was 0.5/9 and for the carcinoma group was 2.8/9. Mean TCIS for the carcinoma group was significantly higher than that of the control group ($p = 0.0001$).

Of all subtypes, tubular adenocarcinoma had the highest TCIS, 3.4, followed by mucinous—3.1, SRC—2.6, and undifferentiated carcinoma—2.5. Papillary adenocarcinoma had the lowest TCIS of 2.0. All subtypes (except for papillary) had a consistently significantly greater TCIS when compared with the control group ($p = 0.0001$) (Figure 3).

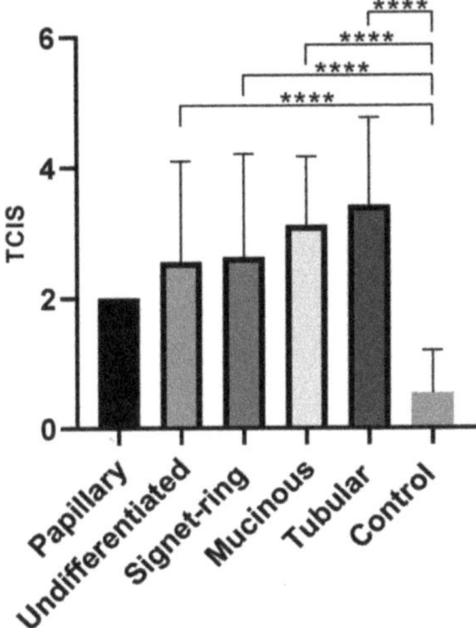

Figure 3. Mean total chronic inflammation score (TCIS) by carcinoma histological subtype. **** $p < 0.0001$.

Fifty out of 149 (33.5%) gastric carcinoma samples had observable *Helicobacter* spp., of which 19 (38%) had large numbers present. The TCIS of *Helicobacter* spp. positive samples was 3.2, versus 2.6 for *Helicobacter* spp. negative samples. These two means were found to be significantly different ($p = 0.039$) (Figure 4).

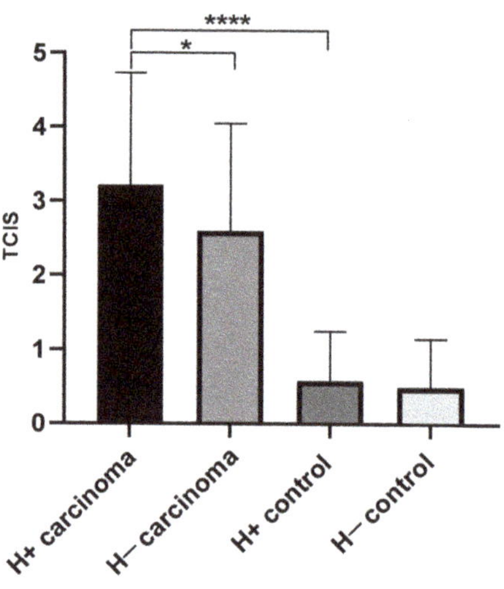

Figure 4. Mean total chronic inflammation score (TCIS) by helicobacter status for carcinoma and control groups. H+, helicobacter positive samples, H−, helicobacter negative samples.

Thirteen out of 29 (44.8%) of the control samples had observable *Helicobacter* spp., but only 2 (15.4%) had large numbers present. The TCIS of *Helicobacter* spp. positive samples was 0.58, and helicobacter negative samples was 0.50. These two means were not significantly different.

3.4. Immunohistochemical Assessment of E-Cadherin, CD44, 14-3-3σ and p16

Twenty-two cases of gastric carcinoma, randomly selected from each histopathological subtype, were available for immunohistochemical assessment as follows: SRC (8, 36.4%), undifferentiated (7, 31.8%), tubular (5, 22.7%), papillary (1, 4.5%), and mucinous (1, 4.5%).

3.5. E-Cadherin Expression

In normal gastric surface and glandular epithelium, E-cadherin labeling was strong (3/3 intensity score) and membranous (Figure 5a). Seventeen (of 22) tumors demonstrated positive immunolabeling for E-cadherin, with both membranous and cytoplasmic (17/22, 77.3%) and nuclear (2/22, 9.1%) labeling, and with an intensity greatest in intravascular emboli, where present. Five tumors did not show immunopositivity for E-cadherin (2 undifferentiated, 1 SRC, 1 mucinous and 1 tubular carcinoma). Where dysplastic epithelium was present, E-cadherin expression decreased in intensity and distribution (compared with normal), and labeling became progressively cytoplasmic (from membranous) (Figure 5b). The TIS E-cadherin for each tumor subtype was as follows: papillary—6.0; undifferentiated—4.6; SRC—4.3; tubular—3.2; and mucinous—0 (Table 2). Aberrantly increased expression and TIS of E-cadherin in intralymphatic/intravascular emboli, compared with both normal epithelium and immunopositive neoplastic cells, was observed in seven tumors (7/17; 3/7 undifferentiated, 1/8 signet-ring cell and 3/5 tubular carcinoma). E-cadherin was not expressed by neoplastic emboli composed of well-differentiated acinar structures (Figure 5c).

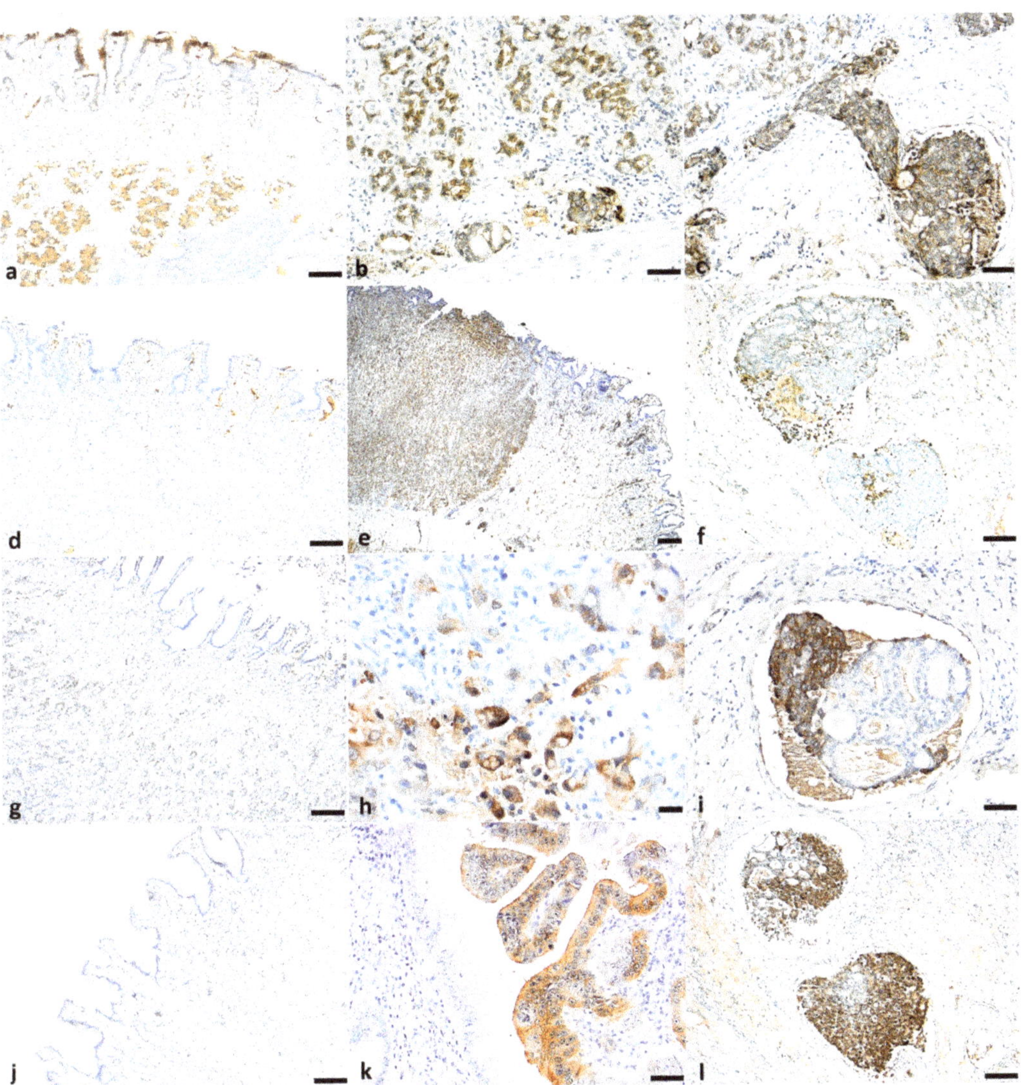

Figure 5. Canine gastric carcinoma immunohistochemistry for E-cadherin, CD44, 14-3-3σ, p16. (**a**) Expression of E-cadherin in normal gastric surface and glandular epithelium (bar = 100 μm). (**b**) Reduced E-cadherin expression (progressively cytoplasmic from membranous) in neoplastic tubules in a tubular adenocarcinoma (bar = 50 μm). (**c**) Poorly differentiated neoplastic cells in emboli showing cytoplasmic and membranous labeling for E-cadherin (bar = 50 μm). (**d**) Expression of CD44 in lymphocytes, macrophages and dendritic cells of normal gastric mucosa. (bar = 100 μm) (**e**) Strong membranous labeling of CD44 in sheets and chains of poorly differentiated neoplastic cells in an undifferentiated gastric carcinoma (bar = 200 μm). (**f**) Loss of membranous and cytoplasmic labeling of CD44 in tubules and strong membranous and cytoplasmic labeling in poorly differentiated cells of a neoplastic embolus (bar = 100 μm). (**g**) Normal epithelium showing no expression of 14-3-3σ (bar = 100 μm). (**h**) Strong cytoplasmic with occasional nuclear labeling of 14-3-3σ in nests of poorly differentiated neoplastic cells in an undifferentiated gastric carcinoma (bar = 20 μm). (**i**) Loss of cytoplasmic labeling of 14-3-3σ in tubules and cytoplasmic labeling in poorly differentiated cells of a neoplastic embolus (bar = 50 μm). (**j**) Normal epithelium showing no expression of p16 (bar = 100 μm). (**k**) Strong cytoplasmic labeling of p16 in neoplastic epithelium (bar = 50 μm). (**l**) Loss of cytoplasmic labeling for p16 in tubules and strong nuclear and cytoplasmic labeling in poorly differentiated cells of a neoplastic embolus (bar = 50 μm).

Table 2. Mean total immunohistochemical score by gastric carcinoma subtype. The same score for neoplastic vascular emboli, where present, is given in parentheses.

Protein	Canine Gastric Carcinoma Subtype				
	Undifferentiated	**Signet-Ring**	**Mucinous**	**Tubular**	**Papillary**
E-cadherin	4.6 (9)	4.3 (6)	0	3.2 (10)	6
CD44	4.7 (6.3)	3 (6.3)	4	2.4 (3.6)	6
14-3-3σ	3.4 (9)	0	0	2.4 (9)	0
p16	4.3 (9)	6.9	9	6.8 (10.5)	6

3.6. CD44 Expression

In the normal canine stomach, CD44 was expressed on the membrane of lymphocytes, macrophages, dendritic cells, and satellite glial cells (Figure 5d). Gastric epithelium, stromal tissue, smooth muscle fibers and matrix fibroblasts were negative. Positive immunolabeling (neo-expression) occurred in all gastric carcinoma cases. CD44 expression was cytoplasmic and membranous (5/22, 22.7%) or only membranous (9/22, 40.9%), with an intensity greatest in mucinous tumors (mean intensity score 2.5). Enhanced membranous expression was observed in those neoplastic epithelial cells that showed the greatest features of malignancy (pleomorphism and invasion), with a higher mean score in papillary, undifferentiated and tubular carcinomas (6, 4.7 and 4, respectively) (Figure 5e). Intravascular neoplastic emboli, present in 4/7 undifferentiated, 4/5 tubular and 3/7 SRC carcinomas, showed increased TIS compared to immunopositive neoplastic cells. Neoplastic cells in emboli exhibited membranous, or both cytoplasmic and membranous expression, in poorly differentiated neoplastic cells. Interestingly, there were neoplastic emboli that had solid nests and well-differentiated acinar arrangements, and, in these emboli, only the solid nests were strongly positive, while the well-differentiated embolic acini were negative (Figure 5f).

3.7. 14-3-3σ Expression

In a histologically normal canine stomach, 14-3-3σ was not expressed in any cell types, including epithelium, stromal tissue or lymphocytes (Figure 5g). Positive immunolabeling (neo-expression) occurred in 10 of 22 (45.4%) gastric carcinomas, 7/7 undifferentiated, and 3/5 tubular. TIS 14-3-3σ for undifferentiated and tubular carcinomas was 3.4 and 2.4, respectively. SRC, papillary and mucinous carcinomas were all negative for 14-3-3σ. Cytoplasmic and/or nuclear neo-expression of 14-3-3σ was present, with an intensity greatest in undifferentiated tumors (intensity mean 2.1, and total score 3.4). One case of undifferentiated carcinoma showed both nuclear and cytoplasmic labeling in neoplastic cells (Figure 5h). Where neoplastic emboli were present, TIS was higher compared to immunopositive neoplastic cells, and 14-3-3σ labeling intensity was increased in poorly differentiated and pleomorphic cells. However, when neoplastic cells formed intravascular acinar structures, 14-3-3σ was not expressed (Figure 5i).

3.8. p16 Expression

In a histologically normal canine stomach, p16 was not expressed (Figure 5j), but positive cytoplasmic and/or nuclear labeling was noted in regions of dysplastic epithelium (Figure 5k). p16 expression occurred in 19/22 (86.4%) carcinomas; 6/7 undifferentiated, 5/5 tubular, 1/1 papillary, 6/8 SRC, and 1/1 mucinous. Positive immunolabeling of pleomorphic cells and dysplastic tubules was found in all tumors, with high-grade expression in mucinous tumors (signet-ring cells in SRC and mucinous carcinomas were strongly immunopositive for p16). p16 exhibited cytoplasmic (11/22, 50%) and both nuclear/cytoplasmic (8/22, 36.36%) expression, with intensity greatest in mucinous tumors (intensity mean 3 and total score 9).

Where neoplastic emboli were present, TIS was higher compared to immunopositive neoplastic cells. Within intravascular and intralymphatic emboli, p16 intensity and cytoplasmic and/or nuclear expression was increased in poorly differentiated and pleomorphic cells. However, when neoplastic cells formed intravascular acinar structures, p16 was not expressed (Figure 5l). Similar findings were noted in the lymph node metastasis of one sample. The mean TIS p16 for undifferentiated, tubular, papillary, SRC and mucinous carcinomas was 4.3, 6.8, 6, 6.9 and 9, respectively.

4. Discussion

In this canine gastric carcinoma (CGC) study, using the WHO scheme, we have demonstrated that the most frequent subtype is the undifferentiated carcinoma, and the most common breed affected is the Staffordshire bull terrier. There is no sex predisposition, and the mean age is nine years. *Helicobacter* spp. presence was associated with increased chronic inflammation parameters and a greater chronic inflammatory score. We found that all markers showed variable expression for each subtype. CD44 and 14-3-3σ have not been previously investigated in CGC. 14-3-3σ was positive in undifferentiated, tubular and papillary carcinomas, and p16 expression was increased in mucinous and SRC carcinomas. E-cadherin and CD44 were variably expressed in all subtypes and were associated with criteria of malignancy. Within neoplastic emboli, acinar structures lacked expression of all markers, suggesting an independent molecular pathway that requires further investigation.

Of the 149 dogs included in this study from the UK and Spain, the mean and median age of 9.1 and 9 years, respectively, at the time of CGC diagnosis is in broad agreement with previously reported studies [3,6,7]. In this study, CGC affected male and female dogs roughly equally, with a slight preponderance towards males, as is consistent with previous reports [3]. The most commonly affected breed was the Staffordshire bull terrier, likely reflecting both the general population in the United Kingdom (only 2/22 Staffordshire bull terriers were from Spain) and previous reports of breed predisposition [7]. Breed predisposition to CGC is also reported in less common breeds, including Belgian shepherd dogs and rough collies, which also appeared with an increased frequency in this study [3,6].

Using the WHO classification for domestic animals, cases were divided into five categories based on the predominant histological features and the principal cell type of the tumor. All histological subtypes of carcinoma (SRC, tubular, mucinous, papillary and undifferentiated) were recorded, although squamous cell carcinoma, another CGC subtype, was not present in this study [44]. Undifferentiated carcinoma was the most frequent subtype, followed by SRC, contrary to the previous studies reporting an increased frequency for the tubular subtype, similar to humans [45,46]. This could likely reflect a geographic variation in gastric adenocarcinoma incidence.

Signet-ring cell carcinomas were the most frequently diagnosed type by endoscopy, whereas undifferentiated carcinomas were the most frequently diagnosed in full-thickness samples. The most likely reason behind this difference is that, histologically, the diagnosis of SRC subtype is based upon identification of the characteristic isolated or small groups of malignant cells containing intracytoplasmic mucin with an eccentric nucleus (signet-ring cells) within the mucosa, and hence, diagnosis from an endoscopically retrieved (mucosa only) sample is possible. The diagnosis of other histopathological subtypes typically requires the evaluation of invasion beyond the mucosa, and thus, given the relatively high number of full-thickness biopsies in this study, this may have influenced the predominant subtype. Additionally, signet-ring cells were also present in other subtypes, i.e., mucinous and undifferentiated carcinoma; however, the predominant histological features and pattern did not favor a diagnosis of SRC carcinoma. This suggests that endoscopic samples alone can lead to diagnostic pitfalls between different subtypes [47–50]. In humans, novel techniques such as endoscopic surgical dissection are recommended in subtypes like SRC that spread subepithelially in the margins [51]. In combination with future novel imaging techniques, these findings should be taken into consideration to determine the appropriate sampling and therapeutic approach [52].

Cancer-related inflammation is one of the hallmarks of neoplasia, and once cancer develops, there is evidence of a substantial shift in the microenvironment affecting the immune response [53,54]. A simplified histopathologic scoring system was adapted to reduce variability in the diagnostic interpretation [37]. This scoring system captured and quantified chronic inflammatory changes in the gastric mucosa [55]. The TCIS for the carcinoma group, and individually for each CGC subtype (independent of helicobacter presence), was significantly greater when compared to the control group's TCIS. Thus, there is a clear correlation between chronic inflammation and carcinoma for most CGC subtypes.

Given the link between the presence of helicobacter, inflammation and neoplasia in humans we aimed to quantitatively assess helicobacter presence in CGCs. In previous studies *Helicobacter* spp. have been reported in both neoplastic and non-neoplastic canine stomachs [56,57]. In this study, the TCIS of the carcinoma group with a concomitant presence of helicobacter was significantly greater compared to the helicobacter-negative carcinoma group. Furthermore, larger numbers of helicobacter were found in carcinoma cases than in control cases. Although the helicobacter presence does not prove a definite role in CGC pathogenesis, there is a clear association between increased chronic inflammation and higher numbers of helicobacter in the tumor cases versus controls. Whether the presence of *Helicobacter* spp. is associated with tumor development is not exactly clear, and thus further studies, perhaps to examine particular species of helicobacter in association with CGC, would be needed.

To understand the molecular pathways and investigate the cellular origin of CGC, the expression of cellular adhesion molecules (E-cadherin and CD44) and cell cycle regulators (p16 and 14-3-3σ) were investigated in a representative proportion of cases. CD44 and 14-3-3σ have not previously been studied in CGC.

E-cadherin controls cell motility and suppresses tumor growth and metastasis [58]. Immunohistochemical examination of E-cadherin in all tumor subtypes identified abnormalities of expression and localization. Dysplastic surface and glandular epithelia in immunopositive cases revealed a reduction in E-cadherin expression. Pleomorphic cells in the undifferentiated subtype showed E-cadherin expression, albeit in decreased intensity compared to the normal epithelium. Subcellular localization of E-cadherin was observed in the majority of neoplastic cells. Intracytoplasmic sequestration and the accumulation of E-cadherin have been previously associated with abnormalities in the intracytoplasmic transport and reuptake mechanisms of the molecule [59,60]. Aberrant nuclear expression of E-cadherin in humans has been previously described in several tumor types including gastric and colorectal carcinomas [19,61]. Similarly, increased intensity intracytoplasmic and nuclear expression was noted in the neoplastic emboli composed of pleomorphic neoplastic cells forming solid nests. Surprisingly, intravascular acinar structures did not express E-cadherin. Furthermore, tubular adenocarcinoma, which is composed of acini and tubules, had the lowest TIS compared to other subtypes. This could represent an inverse association between the reduction or loss of E-cadherin with an increasing degree of differentiation, where well-differentiated structures switch off and pleomorphic cells switch on or increase the expression of E-cadherin. Previous studies in human gastric adenocarcinoma showed that abnormalities of E-cadherin localization (internalization) in neoplastic cells lead to decreased adhesion, thus favoring invasion [59,60]. Neoplastic cells lacked E-cadherin expression in discohesive subtypes like mucinous adenocarcinoma, which is in accordance with a recent study [20].

In the context of these findings, we postulate that the reduction or loss of E-cadherin expression on the cell membrane could potentially facilitate invasion, and that the re-expression of E-cadherin, including on the cell membrane, in neoplastic emboli could lead to the development of solid cohesive intravascular structures, enhancing their survival. These findings reinforce the notion that E-cadherin in CGC is a context-dependent adhesion molecule that can either be up- or down-regulated or re-expressed, depending on the stage of tumor progression [62]. Further studies analyzing ligands of E-cadherin and possible mutations are required to clarify its precise role in CGC.

CD44 serves as a signaling platform, with transmembrane and cytoplasmic domains as a co-receptor for various types of cell surface receptors that modulate cell adhesion, migration, proliferation, differentiation and survival [63–66]. In humans, its role in the adaptive plasticity and survival of cancer cells in processes like epithelial to mesenchymal transition, invasion and metastasis has been widely studied [67,68]. In CGC, CD44 was variably expressed in all tumor subtypes. Similar to E-cadherin, the expression of CD44 was enhanced in pleomorphic neoplastic cells and was absent in well-differentiated acinar structures within neoplastic emboli. In the context of these findings, we postulate that CD44 expression could potentially facilitate invasion and expression in neoplastic emboli and, alongside E-cadherin, could lead to the development of cohesive intravascular structures, enhancing their survival. In previous studies in humans, CD44 positive cells in gastric cell lines were associated with increased chemoresistance and invasiveness [66]. Additionally, CD44 positive gastric tumors were associated with larger tumor size, a lower grade of differentiation, tumor relapse, lymph node invasion, distant metastasis and reduced survival [63–66]. It seems that CD44 may represent an important biomarker and a promising therapeutic target in canine gastric carcinomas.

14-3-3σ is a cell cycle regulator that may serve as either a tumor suppressor or an oncogene involved in tissue invasion and metastasis. Histological features of malignancy have been previously associated with the overexpression and/or neo-expression of the protein [69,70]. The role of this protein as a tissue differentiation marker and as an oncoprotein in veterinary medicine has been previously studied in canine mammary, urinary bladder, renal cell and equine penile squamous cell carcinomas [33,34,71,72]. In the current study, immunohistochemical analysis revealed the absence of 14-3-3σ immunolabeling in a normal stomach and in SRC and mucinous carcinomas. The latter finding was consistent for all cases of the two subtypes and likely demonstrates a 14-3-3σ independent molecular pathway in carcinogenesis. In contrast, tubular and undifferentiated subtypes showed strong intracytoplasmic neoexpression, and occasionally nuclear expression of 14-3-3σ. The single papillary carcinoma had weak intracytoplasmic neoexpression. Undifferentiated carcinomas demonstrated strong intracytoplasmic, and occasionally nuclear, expression in the most pleomorphic neoplastic cells. Similarly, in those tumors with neoplastic emboli formed by pleomorphic cells forming solid nests, there was strong intracytoplasmic and occasionally nuclear expression of 14-3-3σ. However, acinar structures within the intravascular emboli lacked expression of 14-3-3σ. Aberrant nuclear expression of 14-3-3σ, in a subset of cases of renal cell carcinomas, was associated with a malignant phenotype and shorter survival rate [43]. Thus, we hypothesize, in view of these findings, that expression of 14-3-3σ is associated with features of malignancy, and that neoexpression of the molecule is essential in the stage of intravascular invasion. Expression of the protein in the extracellular milieu, as previously demonstrated for other canine carcinomas, was not observed in this study [72].

The protein p16 acts as a tumor suppressor and a cell cycle regulator, by slowing progression from the G1 to S phase through the inhibition of cyclinD-CDK4/6. Its role and possible relationship to tumor progression and prognosis have been studied in a variety of human tumors [73–75]. In human gastric carcinomas, loss of p16 expression has been associated with malignant characteristics and poor prognosis [28,76,77]. In the CGCs studied here, the expression of p16 seemed to be an event common to all subtypes, and thus appears fundamental for neoplasia development, with expression most strong in mucinous and SRC carcinomas. Half of the tumors showed cytoplasmic, and 36.7% both nuclear and cytoplasmic, immunolabeling. Within neoplastic emboli, expression was cytoplasmic and only in one tubular carcinoma case was both cytoplasmic and nuclear. Similar to the other markers, when acinar structures were present within the emboli, p16 expression was absent. These results contradict a previous canine study, where lack of p16 expression in tubular, SRC and undifferenced subtypes was significantly associated with histological criteria of malignancy in 14 cases [29]. In humans, the overexpression of p16 is associated with mutations in genes encoding retinoblastoma protein (Rb) and p53, and is considered to be a mechanism to arrest the uncontrolled proliferation caused by failure of the Rb pathway [78]. Furthermore, the significance of the p16 expression within the cell is not clear, and few studies have associated cytoplasmic expression with malignant features [79]. Currently, similar to humans, the significance of p16 expression patterns in CGC remains unclear [78].

5. Conclusions

In conclusion, CGC affects female and male dogs in relatively equal proportions, with a mean age of nine years, and the highest frequency in Staffordshire bull terriers. The most common CGC histological subtype was the undifferentiated carcinoma. CGCs were associated with increased chronic inflammation parameters and with a greater chronic inflammatory score when *Helicobacter* spp. were present. Understanding the altered molecular pathways associated with gastric carcinoma development including its histological subtype remains a challenge. The significance of the findings from the cell cycle regulators and cell adhesion molecules in the subtypes examined, should be combined with clinicopathological data and additional molecular analysis to further understand the molecular mechanisms and similarities between dogs and humans.

Supplementary Materials: The following are available online at https://www.mdpi.com/article/10.3390/ani11051409/s1, Table S1: 149 cases of canine gastric carcinoma by sex and breed frequency.

Author Contributions: Conceptualization, S.L.P., A.S.-B. and A.H.; methodology, S.L.P., A.S.-B. and A.H.; investigation, A.H. and W.E.B.; resources, S.B. and G.A.R.; data curation, A.H. and W.E.B.; writing—original draft preparation, A.H.; writing—review and editing, A.H., S.L.P. and A.S.-B. All authors have read and agreed to the published version of the manuscript.

Funding: This research received no external funding.

Institutional Review Board Statement: Not applicable.

Data Availability Statement: The data presented in this study is contained within the manuscript and Supplementary Table S1.

Conflicts of Interest: The authors declare no conflict of interest.

References

1. Patnaik, A.K.; Hurvitz, A.I.; Johnson, G.F. Canine gastrointestinal neoplasms. *Vet. Pathol.* **1977**, *14*, 547–555. [CrossRef] [PubMed]
2. Munday, J.S.; Löhr, C.V.; Kiupel, M. Tumors of the alimentary tract. *Tumors Domest. Anim.* **2016**, 499–601. [CrossRef]
3. Fonda, D.; Gualtieri, M.; Scanziani, E. Gastric-carcinoma in the dog—A clinicopathological study of 11 cases. *J. Small Anim. Pract.* **1989**, *30*, 353–360. [CrossRef]
4. Crow, S.E. Tumors of the alimentary tract. *Vet Clin. N. Am. Small Anim. Pract.* **1985**, *15*, 577–596. [CrossRef]
5. De Stefani, E.; Correa, P.; Boffetta, P.; Deneo-Pellegrini, H.; Ronco, A.L.; Mendilaharsu, M. Dietary patterns and risk of gastric cancer: A case-control study in uruguay. *Gastric Cancer Off. J. Int. Gastric Cancer Assoc. Jpn. Gastric Cancer Assoc.* **2004**, *7*, 211–220. [CrossRef]
6. Seim-Wikse, T.; Jorundsson, E.; Nodtvedt, A.; Grotmol, T.; Bjornvad, C.R.; Kristensen, A.T.; Skancke, E. Breed predisposition to canine gastric carcinoma—A study based on the norwegian canine cancer register. *Acta. Vet. Scand.* **2013**, *55*, 25. [CrossRef]
7. Sullivan, M.; Lee, R.; Fisher, E.W.; Nash, A.S.; McCandlish, I.A. A study of 31 cases of gastric carcinoma in dogs. *Vet. Rec.* **1987**, *120*, 79–83. [CrossRef] [PubMed]
8. Head, K. Tumors of the alimentary system of domestic animals. *WHO Collab. Cent. Worldw. Ref. Comp. Oncol.* **2003**, 54–55.
9. Lauren, P. The two histological main types of gastric carcinoma: Diffuse and so-called intestinal-type carcinoma: An attempt at a histo-clinical classification. *Acta Pathol. Microbiol. Scand.* **1965**, *64*, 31–49. [CrossRef]
10. Morrison, W.B. Inflammation and cancer: A comparative view. *J. Vet. Intern. Med.* **2012**, *26*, 18–31. [CrossRef] [PubMed]
11. Shacter, E.; Weitzman, S.A. Chronic inflammation and cancer. *Oncology (Williston Park)* **2002**, *16*, 217–226. [PubMed]
12. Piazuelo, M.B.; Epplein, M.; Correa, P. Gastric cancer: An infectious disease. *Infect. Dis. Clin. N. Am.* **2010**, *24*, 853–869. [CrossRef] [PubMed]
13. Hatakeyama, M. Structure and function of helicobacter pylori caga, the first-identified bacterial protein involved in human cancer. *Proc. Jpn. Acad. Ser. B Phys. Biol. Sci.* **2017**, *93*, 196–219. [CrossRef] [PubMed]
14. Amorim, I.; Smet, A.; Alves, O.; Teixeira, S.; Saraiva, A.L.; Taulescu, M.; Reis, C.; Haesebrouck, F.; Gartner, F. Presence and significance of helicobacter spp. In the gastric mucosa of portuguese dogs. *Gut Pathog.* **2015**, *7*, 12. [CrossRef]
15. Priestnall, S.L.; Wiinberg, B.; Spohr, A.; Neuhaus, B.; Kuffer, M.; Wiedmann, M.; Simpson, K.W. Evaluation of "helicobacter heilmannii" subtypes in the gastric mucosas of cats and dogs. *J. Clin. Microbiol.* **2004**, *42*, 2144–2151. [CrossRef]

16. Rossi, G.; Rossi, M.; Vitali, C.G.; Fortuna, D.; Burroni, D.; Pancotto, L.; Capecchi, S.; Sozzi, S.; Renzoni, G.; Braca, G.; et al. A conventional beagle dog model for acute and chronic infection with helicobacter pylori. *Infect. Immun.* **1999**, *67*, 3112–3120. [CrossRef]

17. Pugacheva, E.N.; Roegiers, F.; Golemis, E.A. Interdependence of cell attachment and cell cycle signaling. *Curr. Opin. Cell Biol.* **2006**, *18*, 507–515. [CrossRef] [PubMed]

18. Berx, G.; van Roy, F. Involvement of members of the cadherin superfamily in cancer. *Cold Spring Harb. Perspect. Biol.* **2009**, *1*, a003129. [CrossRef]

19. Rodriguez, F.J.; Lewis-Tuffin, L.J.; Anastasiadis, P.Z. E-cadherin's dark side: Possible role in tumor progression. *Biochim. Biophys. Acta* **2012**, *1826*, 23–31. [CrossRef]

20. Saito, T.; Chambers, J.K.; Nakashima, K.; Nibe, K.; Ohno, K.; Tsujimoto, H.; Uchida, K.; Nakayama, H. Immunohistochemical analysis of beta-catenin, e-cadherin and p53 in canine gastrointestinal epithelial tumors. *J. Vet. Med. Sci.* **2020**, *82*, 1277–1286. [CrossRef]

21. Aruffo, A.; Stamenkovic, I.; Melnick, M.; Underhill, C.B.; Seed, B. Cd44 is the principal cell surface receptor for hyaluronate. *Cell* **1990**, *61*, 1303–1313. [CrossRef]

22. Orian-Rousseau, V.; Sleeman, J. Cd44 is a multidomain signaling platform that integrates extracellular matrix cues with growth factor and cytokine signals. In *Advances in Cancer Research*; Elsevier: Amsterdam, The Netherlands, 2014; Volume 123, pp. 231–254.

23. Yan, Y.; Zuo, X.; Wei, D. Concise review: Emerging role of cd44 in cancer stem cells: A promising biomarker and therapeutic target. *Stem Cells Transl. Med.* **2015**, *4*, 1033–1043. [CrossRef] [PubMed]

24. Da Cunha, C.B.; Oliveira, C.; Wen, X.; Gomes, B.; Sousa, S.; Suriano, G.; Grellier, M.; Huntsman, D.G.; Carneiro, F.; Granja, P.L.; et al. De novo expression of cd44 variants in sporadic and hereditary gastric cancer. *Lab. Investig.* **2010**, *90*, 1604–1614. [CrossRef]

25. Li, H.; Guo, L.; Li, J.W.; Liu, N.; Qi, R.; Liu, J. Expression of hyaluronan receptors cd44 and rhamm in stomach cancers: Relevance with tumor progression. *Int. J. Oncol.* **2000**, *17*, 927–932. [CrossRef]

26. Li, M.; Zhang, B.; Zhang, Z.; Liu, X.; Qi, X.; Zhao, J.; Jiang, Y.; Zhai, H.; Ji, Y.; Luo, D. Stem cell-like circulating tumor cells indicate poor prognosis in gastric cancer. *Biomed. Res. Int.* **2014**, *2014*, 981261. [CrossRef]

27. Fang, M.; Wu, J.; Lai, X.; Ai, H.; Tao, Y.; Zhu, B.; Huang, L. Cd44 and cd44v6 are correlated with gastric cancer progression and poor patient prognosis: Evidence from 42 studies. *Cell Physiol. Biochem.* **2016**, *40*, 567–578. [CrossRef] [PubMed]

28. Myung, N.; Kim, M.R.; Chung, I.P.; Kim, H.; Jang, J.J. Loss of p16 and p27 is associated with progression of human gastric cancer. *Cancer. Lett.* **2000**, *153*, 129–136. [CrossRef]

29. Carrasco, V.; Canfran, S.; Rodriguez-Franco, F.; Benito, A.; Sainz, A.; Rodriguez-Bertos, A. Canine gastric carcinoma: Immunohistochemical expression of cell cycle proteins (p53, p21, and p16) and heat shock proteins (hsp27 and hsp70). *Vet. Pathol.* **2011**, *48*, 322–329. [CrossRef]

30. Muhlmann, G.; Ofner, D.; Zitt, M.; Muller, H.M.; Maier, H.; Moser, P.; Schmid, K.W.; Zitt, M.; Amberger, A. 14-3-3 sigma and p53 expression in gastric cancer and its clinical applications. *Dis. Markers* **2010**, *29*, 21–29. [CrossRef] [PubMed]

31. Li, Y.L.; Liu, L.; Xiao, Y.; Zeng, T.; Zeng, C. 14-3-3σ is an independent prognostic biomarker for gastric cancer and is associated with apoptosis and proliferation in gastric cancer. *Oncol. Lett.* **2015**, *9*, 290–294. [CrossRef] [PubMed]

32. Lodygin, D.; Hermeking, H. Epigenetic silencing of 14-3-3sigma in cancer. In *Seminars in Cancer Biology*; Elsevier: Amsterdam, The Netherlands, 2006; pp. 214–224.

33. Suarez-Bonnet, A.; Herraez, P.; de las Mulas, J.M.; Rodriguez, F.; Deniz, J.M.; de los Monteros, A.E. Expression of 14-3-3 sigma protein in normal and neoplastic canine mammary gland. *Vet. J.* **2011**, *190*, 345–351. [CrossRef] [PubMed]

34. Suárez-Bonnet, A.; Herráez, P.; Aguirre, M.; Suárez-Bonnet, E.; Andrada, M.; Rodríguez, F.; de los Monteros, A.E. Expression of cell cycle regulators, 14-3-3σ and p53 proteins, and vimentin in canine transitional cell carcinoma of the urinary bladder. *Urol. Oncol. Semin. Orig. Investig.* **2015**, *33*, 332.e1.

35. Fenoglio-Preiser, C.; Carneiro, F.; Correa, P.; Guilford, P.; Lambert, R.; Megraud, F.; Muñoz, N.; Powell, S.M.; Rugge, M.; Sasako, M.; et al. Tumours of the stomach. In *World Health Organization Classification of Tumours. Pathology and Genetics of Tumours of the Digestive System*; Hamilton, S., Aaltonen, L., Eds.; IARC Press: Lyon, France, 2000; pp. 37–67.

36. Head, K.W.; Armed Forces Institute of Pathology (U.S.); American Registry of Pathology; WHO Collaborating Center for Worldwide Reference on Comparative Oncology. *Histological Classification of Tumors of the Alimentary System of Domestic Animals*; International Histological Classification of Tumors of Domestic Animals Series; Armed Forces Institute of Pathology: Washington, DC, USA, 2003.

37. Allenspach, K.A.; Mochel, J.P.; Du, Y.; Priestnall, S.L.; Moore, F.; Slayter, M.; Rodrigues, A.; Ackermann, M.; Krockenberger, M.; Mansell, J.; et al. Correlating gastrointestinal histopathologic changes to clinical disease activity in dogs with idiopathic inflammatory bowel disease. *Vet. Pathol.* **2019**, *56*, 435–443. [CrossRef]

38. Day, M.J.; Bilzer, T.; Mansell, J.; Wilcock, B.; Hall, E.J.; Jergens, A.; Minami, T.; Willard, M.; Washabau, R. Histopathological standards for the diagnosis of gastrointestinal inflammation in endoscopic biopsy samples from the dog and cat: A report from the world small animal veterinary association gastrointestinal standardization group. *J. Comp. Pathol.* **2008**, *138*, S1–S43. [CrossRef]

39. Querzoli, P.; Coradini, D.; Pedriali, M.; Boracchi, P.; Ambrogi, F.; Raimondi, E.; La Sorda, R.; Lattanzio, R.; Rinaldi, R.; Lunardi, M. An immunohistochemically positive e-cadherin status is not always predictive for a good prognosis in human breast cancer. *Br. J. Cancer* **2010**, *103*, 1835–1839. [CrossRef] [PubMed]

40. Da Cunha, I.W.; Souza, M.J.L.; da Costa, W.H.; Amâncio, A.M.; Fonseca, F.P.; de Cassio Zequi, S.; Lopes, A.; Guimarães, G.C.; Soares, F. Epithelial-mesenchymal transition (emt) phenotype at invasion front of squamous cell carcinoma of the penis influences oncological outcomes. *Urol. Oncol. Semin. Orig. Investig.* **2016**, *34*, 433.e19.
41. Rogez, B.; Pascal, Q.; Bobillier, A.; Machuron, F.; Lagadec, C.; Tierny, D.; Le Bourhis, X.; Chopin, V. Cd44 and cd24 expression and prognostic significance in canine mammary tumors. *Vet. Pathol.* **2019**, *56*, 377–388. [CrossRef] [PubMed]
42. Roma, A.A.; Goldblum, J.R.; Fazio, V.; Yang, B. Expression of 14-3-3σ, p16 and p53 proteins in anal squamous intraepithelial neoplasm and squamous cell carcinoma. *Int. J. Clin. Exp. Pathol.* **2008**, *1*, 419. [PubMed]
43. Suárez-Bonnet, A.; Lara-García, A.; Stoll, A.L.; Carvalho, S.; Priestnall, S.L. 14-3-3σ protein expression in canine renal cell carcinomas. *Vet. Pathol.* **2018**, *55*, 233–240. [CrossRef]
44. Patnaik, A.K.; Lieberman, P.H. Gastric squamous cell carcinoma in a dog. *Vet. Pathol.* **1980**, *17*, 250–253. [CrossRef] [PubMed]
45. Washabau, R.; Day, M. *Canine and Feline Gastroenterology-e-Book*; Elsevier Health Sciences: Philadelphia, PA, USA, 2012.
46. Fléjou, J.-F. Classification oms 2010 des tumeurs digestives: La quatrième édition. *Ann. Pathol.* **2011**, *31*, S27–S31. (In French) [CrossRef]
47. Jergens, A.E.; Willard, M.D.; Allenspach, K. Maximizing the diagnostic utility of endoscopic biopsy in dogs and cats with gastrointestinal disease. *Vet. J.* **2016**, *214*, 50–60. [CrossRef]
48. Standards of Practice Comittee; Faulx, A.L.; Kothari, S.; Acosta, R.D.; Agrawal, D.; Bruining, D.H.; Chandrasekhara, V.; Eloubeidi, M.A.; Fanelli, R.D.; Gurudu, S.R.; et al. The role of endoscopy in subepithelial lesions of the gi tract. *Gastrointest. Endosc.* **2017**, *85*, 1117–1132. [CrossRef]
49. Papanikolaou, I.S.; Triantafyllou, K.; Kourikou, A.; Rosch, T. Endoscopic ultrasonography for gastric submucosal lesions. *World J. Gastrointest. Endosc.* **2011**, *3*, 86–94. [CrossRef] [PubMed]
50. Lee, J.; Oh, S.J. Signet ring cell carcinoma mimicking gastric gastrointestinal stromal tumor: A case report. *Case Rep. Oncol.* **2020**, *13*, 538–543. [CrossRef]
51. Kim, H.; Kim, J.-H.; Lee, Y.C.; Kim, H.; Youn, Y.H.; Park, H.; Choi, S.H.; Noh, S.H.; Gotoda, T. Growth patterns of signet ring cell carcinoma of the stomach for endoscopic resection. *Gut Liver* **2015**, *9*, 720. [CrossRef] [PubMed]
52. Tanaka, T.; Akiyoshi, H.; Mie, K.; Okamoto, M.; Yoshida, Y.; Kurokawa, S. Contrast-enhanced computed tomography may be helpful for characterizing and staging canine gastric tumors. *J. Vet. Radiol. Ultrasound* **2019**, *60*, 7–18. [CrossRef]
53. Hanahan, D.; Weinberg, R.A. Hallmarks of cancer: The next generation. *Cell* **2011**, *144*, 646–674. [CrossRef]
54. Macarthur, M.; Hold, G.L.; El-Omar, E.M. Inflammation and cancer ii. Role of chronic inflammation and cytokine gene polymorphisms in the pathogenesis of gastrointestinal malignancy. *Am. J. Physiol. Gastrointest. Liver Physiol.* **2004**, *286*, G515–G520. [CrossRef] [PubMed]
55. Jergens, A.E.; Evans, R.B.; Ackermann, M.; Hostetter, J.; Willard, M.; Mansell, J.; Bilzer, T.; Wilcock, B.; Washabau, R.; Hall, E.J.; et al. Design of a simplified histopathologic model for gastrointestinal inflammation in dogs. *Vet. Pathol.* **2014**, *51*, 946–950. [CrossRef] [PubMed]
56. Poutahidis, T.; Doulberis, M.; Karamanavi, E.; Angelopoulou, K.; Koutinas, C.K.; Papazoglou, L.G. Primary gastric choriocarcinoma in a dog. *J. Comp. Pathol.* **2008**, *139*, 146–150. [CrossRef]
57. Taulescu, M.A.; Valentine, B.A.; Amorim, I.; Gartner, F.; Dumitrascu, D.L.; Gal, A.F.; Sevastre, B.; Catoi, C. Histopathological features of canine spontaneous non-neoplastic gastric polyps—A retrospective study of 15 cases. *Histol. Histopathol.* **2014**, *29*, 65–75.
58. Harris, T.J.; Tepass, U. Adherens junctions: From molecules to morphogenesis. *Nat. Rev. Mol. Cell Biol* **2010**, *11*, 502–514. [CrossRef]
59. Handschuh, G.; Candidus, S.; Luber, B.; Reich, U.; Schott, C.; Oswald, S.; Becke, H.; Hutzler, P.; Birchmeier, W.; Hofler, H.; et al. Tumour-associated e-cadherin mutations alter cellular morphology, decrease cellular adhesion and increase cellular motility. *Oncogene* **1999**, *18*, 4301–4312. [CrossRef]
60. Gabbert, H.E.; Mueller, W.; Schneiders, A.; Meier, S.; Moll, R.; Birchmeier, W.; Hommel, G. Prognostic value of e-cadherin expression in 413 gastric carcinomas. *Int. J. Cancer.* **1996**, *69*, 184–189. [CrossRef]
61. Moon, K.C.; Cho, S.Y.; Lee, H.S.; Jeon, Y.K.; Chung, J.H.; Jung, K.C.; Chung, D.H. Distinct expression patterns of e-cadherin and beta-catenin in signet ring cell carcinoma components of primary pulmonary adenocarcinoma. *Arch. Pathol. Lab. Med.* **2006**, *130*, 1320–1325. [CrossRef] [PubMed]
62. Fearon, E.R. Cancer: Context is key for e-cadherin in invasion and metastasis. *Curr. Biol.* **2019**, *29*, R1140–R1142. [CrossRef] [PubMed]
63. Han, Y.; Lu, S.; Wen, Y.G.; Yu, F.D.; Zhu, X.W.; Qiu, G.Q.; Tang, H.M.; Peng, Z.H.; Zhou, C.Z. Overexpression of hoxa10 promotes gastric cancer cells proliferation and hoxa10(+)/cd44(+) is potential prognostic biomarker for gastric cancer. *Eur. J. Cell Biol.* **2015**, *94*, 642–652. [CrossRef] [PubMed]
64. Isozaki, H.; Ohyama, T.; Mabuchi, H. Expression of cell adhesion molecule cd44 and sialyl lewis a in gastric carcinoma and colorectal carcinoma in association with hepatic metastasis. *Int. J. Oncol.* **1998**, *13*, 935–942. [CrossRef] [PubMed]
65. Watanabe, T.; Okumura, T.; Hirano, K.; Yamaguchi, T.; Sekine, S.; Nagata, T.; Tsukada, K. Circulating tumor cells expressing cancer stem cell marker cd44 as a diagnostic biomarker in patients with gastric cancer. *Oncol. Lett.* **2017**, *13*, 281–288. [CrossRef]

66. Yoon, C.; Park, D.J.; Schmidt, B.; Thomas, N.J.; Lee, H.J.; Kim, T.S.; Janjigian, Y.Y.; Cohen, D.J.; Yoon, S.S. Cd44 expression denotes a subpopulation of gastric cancer cells in which hedgehog signaling promotes chemotherapy resistance. *Clin. Cancer Res.* **2014**, *20*, 3974–3988. [CrossRef]

67. Thorne, R.F.; Legg, J.W.; Isacke, C.M. The role of the cd44 transmembrane and cytoplasmic domains in co-ordinating adhesive and signalling events. *J. Cell Sci.* **2004**, *117*, 373–380. [CrossRef] [PubMed]

68. Orian-Rousseau, V. Cd44 acts as a signaling platform controlling tumor progression and metastasis. *Front. Immunol.* **2015**, *6*, 154. [CrossRef]

69. Ko, S.; Kim, J.Y.; Jeong, J.; Lee, J.E.; Yang, W.I.; Jung, W.H. The role and regulatory mechanism of 14-3-3 sigma in human breast cancer. *J. Breast Cancer* **2014**, *17*, 207–218. [CrossRef] [PubMed]

70. Mikami, T.; Maruyama, S.; Abé, T.; Kobayashi, T.; Yamazaki, M.; Funayama, A.; Shingaki, S.; Kobayashi, T.; Jun, C.; Saku, T. Keratin 17 is co-expressed with 14-3-3 sigma in oral carcinoma in situ and squamous cell carcinoma and modulates cell proliferation and size but not cell migration. *Virchows Arch.* **2015**, *466*, 559–569. [CrossRef]

71. Suárez-Bonnet, A.; de las Mulas, J.M.; Herráez, P.; Rodríguez, F.; de los Monteros, A.E. Immunohistochemical localisation of 14-3-3 σ protein in normal canine tissues. *Vet. J.* **2010**, *185*, 218–221. [CrossRef] [PubMed]

72. Suarez-Bonnet, A.; Willis, C.; Pittaway, R.; Smith, K.; Mair, T.; Priestnall, S.L. Molecular carcinogenesis in equine penile cancer: A potential animal model for human penile cancer. *Urol. Oncol. Semin. Orig. Investig.* **2018**, *36*, 532.e9. [CrossRef] [PubMed]

73. Herman, J.G.; Merlo, A.; Mao, L.; Lapidus, R.G.; Issa, J.P.J.; Davidson, N.E.; Sidransky, D.; Baylin, S.B. Inactivation of the cdkn2/p16/mts1 gene is frequently associated with aberrant DNA methylation in all common human cancers. *Cancer Res.* **1995**, *55*, 4525–4530.

74. Otterson, G.A.; Kratzke, R.A.; Coxon, A.; Kim, Y.W.; Kaye, F.J. Absence of p16(ink4) protein is restricted to the subset of lung-cancer lines that retains wildtype rb. *Oncogene* **1994**, *9*, 3375–3378.

75. Kelley, M.J.; Nakagawa, K.; Steinberg, S.M.; Mulshine, J.L.; Kamb, A.; Johnson, B.E. Differential inactivation of cdkn2 and rb protein in non—Small-cell and small-cell lung cancer cell lines. *J. Natl. Cancer Inst.* **1995**, *87*, 756–761. [CrossRef]

76. Rocco, A.; Schandl, L.; Nardone, G.; Tulassay, Z.; Staibano, S.; Malfertheiner, P.; Ebert, M. Loss of expression of tumor suppressor p16ink4 protein in human primary gastric cancer is related to the grade of differentiation. *J. Dig. Dis.* **2002**, *20*, 102–105. [CrossRef] [PubMed]

77. Wu, J.F.; Shao, J.C.; Wang, D.B.; Qin, R.; Zhang, H. Expression and significance of cell cycle regulators in gastric carcinoma. *Ai Zheng Chin. J. Cancer* **2005**, *24*, 175–179.

78. Romagosa, C.; Simonetti, S.; Lopez-Vicente, L.; Mazo, A.; Lleonart, M.E.; Castellvi, J.; Ramon y Cajal, S. P16(ink4a) overexpression in cancer: A tumor suppressor gene associated with senescence and high-grade tumors. *Oncogene* **2011**, *30*, 2087–2097. [CrossRef] [PubMed]

79. Liu, Q.; Song, L.J.; Xu, W.Q.; Zhao, L.; Zheng, L.; Yan, Z.W.; Fu, G.H. Expression of cytoplasmic p16 and anion exchanger 1 is associated with the invasion and absence of lymph metastasis in gastric carcinoma. *Mol. Med. Rep.* **2009**, *2*, 169–174. [PubMed]

Article

Immunohistological Evaluation of Von Willebrand Factor in the Left Atrial Endocardium and Atrial Thrombi from Cats with Cardiomyopathy

Wan-Ching Cheng [1,*], Lois Wilkie [1], Tsumugi Anne Kurosawa [1], Melanie Dobromylskyj [2], Simon Lawrence Priestnall [3], Virginia Luis Fuentes [1] and David J. Connolly [1]

[1] Department of Clinical Science and Services, Royal Veterinary College, Hawkshead Lane, North Mymms, Hatfield AL9 7TA, UK; lwilkie@rvc.ac.uk (L.W.); tkurosawa@rvc.ac.uk (T.A.K.); vluisfuentes@rvc.ac.uk (V.L.F.); dconnolly@rvc.ac.uk (D.J.C.)

[2] Finn Pathologists, 3c, 3 Mayflower Way, Harleston IP20 9EB, UK; mdobromylskyj@rvc.ac.uk

[3] Department of Pathobiology and Population Sciences, Royal Veterinary College, Hawkshead Lane, North Mymms, Hatfield AL9 7TA, UK; spriestnall@rvc.ac.uk

[*] Correspondence: wccheng@rvc.ac.uk

Citation: Cheng, W.-C.; Wilkie, L.; Kurosawa, T.A.; Dobromylskyj, M.; Priestnall, S.L.; Luis Fuentes, V.; Connolly, D.J. Immunohistological Evaluation of Von Willebrand Factor in the Left Atrial Endocardium and Atrial Thrombi from Cats with Cardiomyopathy. *Animals* **2021**, *11*, 1240. https://doi.org/10.3390/ani11051240

Academic Editor: Lysimachos Papazoglou

Received: 22 March 2021
Accepted: 19 April 2021
Published: 26 April 2021

Publisher's Note: MDPI stays neutral with regard to jurisdictional claims in published maps and institutional affiliations.

Simple Summary: Disease of the heart muscle (cardiomyopathy) is very common in the domestic cat and may result in several severe outcomes. These include formation of a thrombus in the left atrium which migrates to the hindlimb cutting off the blood supply, a condition called aortic thromboembolism. Affected cats present with hindlimb paralysis and extreme pain, often requiring euthanasia on humane grounds. Several factors are known to predispose to thrombus formation, including damage to the inner cellular lining of the atrium which exposes proteins that initiates thrombosis. We studied the expression of one such protein called von Willebrand Factor in the left atrium of cats with and without cardiomyopathies and at different stages of disease severity. We found that expression increased in cats with advance disease. Obtaining a greater understanding of the role this protein has in thrombus formation may allow development of novel antithrombotic agents to help prevent this devastating consequence of feline cardiomyopathy.

Abstract: Aortic thromboembolism (ATE) occurs in cats with cardiomyopathy and often results in euthanasia due to poor prognosis. However, the underlying predisposing mechanisms leading to left atrial (LA) thrombus formation are not fully characterised. von Willebrand Factor (vWF) is a marker of endothelium and shows increased expression following endothelial injury. In people with poor LA function and LA remodelling, vWF has been implicated in the development of LA thrombosis. In this study we have shown (1) the expression of endocardial vWF protein detected using immunohistofluorescence was elevated in cats with cardiomyopathy, LA enlargement (LAE) and clinical signs compared to cats with subclinical cardiomyopathy and control cats; (2) vWF was present at the periphery of microthrombi and macrothrombi within the LA where they come into contact with the LA endocardium and (3) vWF was integral to the structure of the macrothrombi retrieved from the atria. These results provide evidence for damage of the endocardial endothelium in the remodelled LA and support a role for endocardial vWF as a pro-thrombotic substrate potentially contributing to the development of ATE in cats with underlying cardiomyopathy and LAE. Results from this naturally occurring feline model may inform research into human thrombogenesis.

Keywords: von Willebrand factor; cardiomyopathy; endocardium; left atrial enlargement; immunohistochemistry; aortic thromboembolism

1. Introduction

Feline aortic thromboembolism (ATE) is a severe complication that can occur in 11.6–21% of cats with myocardial disease [1–6]. It happens when a thrombus usually originating in the left atrium (LA) or left atrial appendage (LAA) dislodges and obstructs a branching artery of the aorta. Common clinical signs frequently relate to the pelvic limbs and include severe pain, paraparesis, paraplegia, absence or reduced strength of femoral pulses, and cyanotic paws. ATE in cats carries a poor prognosis, with over 60% of affected cats being euthanised on presentation at first opinion practices [7].

Virchow's triad describes three factors that contribute to venous thrombosis: blood stasis, endothelial injury, and hypercoagulability [8]. Abnormal findings associated with Virchow's triad have been reported in cats with a variety of cardiomyopathies, which provides insights into the potential pathogenesis of feline ATE. For instance, development of spontaneous echo contrast (SEC) in the LA of cats with cardiomyopathy is associated with decreased LAA blood velocity and blood stasis [9]. A hypercoagulable state has been suggested in cats with HCM [10] and histopathologic evidence of LA endothelial damage has been observed in cats with CHF [11]. A further indication of endothelial dysfunction in cats with ATE is that plasma arginine, the precursor to nitric oxide critical for endothelial health, is lower in these cats than in than cats with cardiomyopathy alone [12]. However, the contribution of endothelial injury in the left atrial endocardium to the development of ATE has not been studied at the sub-cellular level in cats with cardiomyopathy at different stages of their disease process and only to a limited extent in humans with heart disease.

vWF is synthesised in endothelial cells and megakaryocytes and is an adhesive protein that plays an important role in thrombosis through the formation of multimers [13,14]. Endothelial vWF is either secreted into the plasma constitutively, stored in rod-shaped specialised compartments called Weibel-Palade bodies within the endothelial cells, or deposited in the subendothelium [15]. vWF of platelet origin is stored in alpha-granules within the platelet and is released upon activation. The main functions of vWF are to aid binding of platelets to exposed subendothelium and to assist in platelet aggregation via the glycoprotein Ib receptor on platelets [16,17]. The majority of plasma vWF originates from the endothelium, and in human patients the plasma concentration of vWF increases in various thrombogenic diseases reflecting endothelial injury [18]. Similarly, in cats with ATE and cardiomyopathy, the elevation in circulatory vWF has been suggested to be associated with endothelial damage [10].

In human patients with atrial distension due to a variety of cardiac conditions, expression of circulating and LA endocardial vWF protein is increased and associated with the severity of LA blood stasis and atrial remodelling and has been suggested as a predisposing factor for thrombogenesis [19–23].

Previous studies have also confirmed the pathogenic role of vWF in both arterial and venous thrombosis. High haemodynamic forces present in the systemic arterial system result in platelets binding to vWF through integrin $\alpha IIb\beta 3$ for the initiation of aggregation [13,24], while patients with vWF deficiency as a result of von Willebrand disease are partly protected against arterial thrombosis [25]. vWF also contributes to thrombosis in veins, through formation of an extracellular scaffold with platelets, leukocytes, and fibrin to trap erythrocytes [26]. Mice deficient in vWF were protected from induced deep vein thrombosis [26]. Although fibrin is the best studied factor involved in trapping red blood cells (RBCs) in slow blood velocity environments [27–30], a recent in vitro study showed that vWF can bind to RBCs directly and the degree of binding is increased when blood stasis is present [31].

Currently, there is limited information regarding the contribution of endocardial vWF to thrombosis in cats with myocardial disease. However, given the pro-thrombotic action of vWF in both rodent models and human patients described above, it is likely that vWF also plays an important role in the development of LA thrombosis in cats. The aim of this study was to investigate the expression of vWF in the LA of cats at different clinical stages of myocardial disease.

2. Materials and Methods

2.1. Study Population

Cats with clinical signs related to cardiomyopathy were enrolled based on their clinical presentation of ATE, CHF or both. The control cats and those with preclinical cardiomyopathy (without clinical signs of disease) were recruited from referral and first opinion cases euthanised for a variety of reasons. Cardiomyopathy was confirmed based on characteristic cardiac structural changes on gross and histopathology using criteria previously described [32,33] (see supplementary material Table S1 for histopathological diagnostic criteria for feline cardiomyopathies) in addition to complete or point of care (POC) echocardiographic examination [34–36] (see Table S2 for echocardiographic diagnostic criteria for feline cardiomyopathies). Briefly for full echocardiographic examinations, all measurements were taken over three different cardiac cycles and averaged. Measurements taken included left atrium to aortic ratio (LA/Ao), maximal left ventricular freewall thickness in diastole (LVFWd), maximal interventricular septum thickness in diastole (IVSd), presence of systolic anterior motion of the mitral valve (SAM), presence of spontaneous echo contrast (SEC), and presence of a formed thrombus in the LA. SAM was defined as anterior motion of either septal or both mitral valve leaflets during systole toward the LVOT using the right parasternal long axis five chamber view on review of 2D cineloops [37]. All echocardiographic examinations were performed by a veterinary cardiology diplomate or resident under direct supervision (see supplementary material Table S5 for echocardiographic views used [38,39]). POC examinations were performed in the emergency setting by a veterinary ECC diplomate or resident in training and facilitated measurement of LA/Ao as described above and a subjective assessment of left ventricular wall thickness in diastole.

Cats were divided into four groups: (1) Control group: cats without structural and histopathological cardiac changes, (2) Subclinical group: cats with subclinical cardiomyopathy (some cats had mild LAE), (3) CHF group: cats with CHF attributable to cardiomyopathy with LAE, (4) ATE group: cats with ATE attributable to cardiomyopathy with LAE, irrespective of whether presenting with concurrent CHF.

Inclusion and exclusion criteria are listed in Table 1.

Table 1. Inclusion and exclusion criteria for case selection of cats by pathological examination.

	Control	Subclinical
Inclusion Criteria	Cats that died of non-cardiac disease with no cardiac related abnormalities detected by clinical exam, gross and histopathology ± complete or POC echocardiography.	Cats that showed no clinical signs of heart disease. Cardiomyopathy confirmed on gross and histopathology + complete or POC echocardiography.
	CHF	**ATE**
	Cats that showed clinical signs of CHF. Cardiomyopathy and LAE confirmed on gross pathology and histopathology + complete or POC echocardiography.	Cats that showed clinical signs compatible with ATE (irrespective of CHF). Cardiomyopathy and LAE confirmed on gross pathology and histopathology + complete or POC echocardiography.
	For all cats with cardiomyopathy	
Exclusion Criteria	Documented chronic renal disease with hypertension, hyperthyroidism, and other uncontrolled systemic diseases that may induce structural cardiac changes.	

See supplementary material Tables S1 and S2 for diagnostic criteria for cardiomyopathies by histopathology and echocardiography.

Diagnosis of ATE was based on clinical signs including acute fore/hind limb(s) paresis or plegia, loss of palpable femoral pulses, cold limbs, and cyanotic nail beds in association with myocardial disease identified by cardiac imaging [40]. CHF was diagnosed based on radiographic evidence of cardiogenic pulmonary oedema, ultrasonographical evidence of pleural effusion in association with cardiomyopathy, and left or bi-atrial enlargement. Cats with cardiomyopathy identified by cardiac imaging and gross and histopathology but showing no clinical signs of heart disease were determined to have subclinical cardiomyopathy.

2.2. Sample Collection

The heart was harvested within 30 min of euthanasia and flushed with slowly running tap water and each chamber was further gently flushed using a 20 mL syringe to ensure all blood was removed to facilitate optimal fixation in 10% buffered formalin. The LA samples were collected from the LA free wall. Where a formed thrombus in the LA was identified, it was carefully removed and placed into 10% buffered formalin, while an in situ thrombus in the LAA was harvested without being removed from the LAA. After gross pathological examination, sections of the left atrial free wall and cross-sections of the ventricles at the heart base, midwall, and apex were embedded in paraffin wax in a routine manner for preparation of slides for haematoxylin and eosin and Masson's trichrome staining for histopathological examination. The LA sample blocks were used to prepare slides for immunostaining.

2.3. Fluorescent Immunostaining

After dewaxing, rehydration, antigen retrieval with citric acid (pH6) at 95 °C for 10 min, and blocking with 10% goat serum, slides were incubated with primary antibodies (1:500 Rabbit polyclonal IgG against vWF, Sigma, Gillingham UK; 1:25 Mouse monoclonal IgG1 against CD41 platelet marker integrin αIIb, clone B-9, Santa Cruz, CA, USA) at 4 °C overnight. An hour-long incubation of slides with 1% Bovine serum albumin/Tris-buffered saline (TBS) suspended 4′,6-diamidino-2-phenylindole (DAPI) (1:100 nuclei stain, Sigma) and secondary antibodies (conjugated with Cyanide Dyes, Cy2 1:100, Cy3 1:500, Jackson ImmunoResearch, Ely, UK) was completed at room temperature. IgG isotype control (1:400 Rabbit polyclonal IgG, Abcam, Cambridge, UK) was used to replace primary antibody against vWF to ensure specific binding. For assessment of nonspecific binding of secondary antibodies, reagent control was carried out by omitting all primary antibodies.

2.4. Image Analysis

All the slides were examined under a Leica DMRA2 microscope (Leica, Wetzlar, Germany) connected to a monochrome camera (AxioCam, Oberkochen, Germany) and 3 images covering the entire length of the LA endocardium were taken. The green fluorescence shown in grey scale from endocardial endothelium was selected free-hand and measured using ImageJ (https://imagej.net/Welcome, accessed on 28 February 2018). The three measurements were then averaged to give a final number representing the detected fluorescence of the endocardial sample. Exposure time was fixed for channels that detected fluorescence from DAPI and vWF for all slides. All the slides were coded so the observer (WCC) was fully blinded when analysing the images and the order in which they were viewed was randomised by a second person (DJC). The fluorescent signals from both IgG isotype control and reagent control were graphed for reference.

A Leica DM4000B with DFC550 colour microscopy camera (Leica, Wetzlar, Germany) was used for light microscopy. The microscopes and cameras were controlled using the Leica Application Suite Version 4.12 (Leica, Wetzlar, Germany).

2.5. Statistical Analysis

All statistical analyses were performed on GraphPad (version 8) (GraphPad, San Diego, CA, USA). Histogram and Shapiro-Wilk test were used for inspection of data distribution and normality. ANOVA test with Tukey post-hoc test and Chi-squared test with Yate correction were used to assess the difference in age and sex. Kruskal-Wallis test with Dunn's multiple comparison test was used to compare the fluorescent intensity of labelled vWF between groups. Difference with a p-Value < 0.05 was considered significant.

3. Results

3.1. Animals

LA free wall samples from 39 cats were used for quantification of endocardial vWF. Thrombi were also retrieved from the LA of 3 (out of the total 39 cats) for immunohistological investigation. The LAA thrombus in situ was acquired from an extra cat that was collected later in the timeline thus not included in the quantification of endocardial vWF (Cat ATE 12 in supplementary material Tables S3 and S4). The inclusion and exclusion criteria for the 4 separate groups of cats are shown in Table 1. The cats in the control group were generally younger but this was not statistically significant ($p = 0.137$). Male cats were overrepresented in cats with cardiomyopathy ($p = 0.009$). Further demographic information about the cats used in the study are given in Table 2.

Table 2. Demographics of the cats enrolled.

Group	Number of Cats	Median Age (Range) (Years)	Gender Male:Female	Breed (Number of Cats)
Control	11	2.0 (0.2–7.5)	4:7	DSH (11)
Subclinical	9	7.5 (2.5–11.0)	6:3	DSH (6) DLH (1) BSH (1) Bengal (1)
CHF	8	7.5 (2.0–17.3)	6:2	DSH (6) BSH (1) Siamese (1)
ATE	11	7 (1.8–11.0)	11:0	DSH (7) DLH (1) BSH (2) Siamese (1)

See supplementary material Tables S3 and S4 for details of clinical presentation, cardiac imaging and histopathological diagnosis for each cat used in this study. One cat in the CHF group (Cat CHF 7 in Tables S3 and S4) had been given clopidogrel prior to presentation.

3.2. Localisation of vWF Protein in the Left Atrial Samples

In all cats, vWF could be observed to a variable degree in the vascular endothelium (Figure 1A,C). In the majority of cats with cardiomyopathy and clinical signs (groups ATE and CHF), microthrombi, defined as thrombi only visible on microscopic examination, were identified on the vascular endothelium and occasionally on the endocardium. Conversely, microthrombi were rarely seen in the control and subclinical groups. Identifiable components of these microthrombi using immunohistofluorescence were vWF (green), platelets (reddish orange), RBC (mild autofluorescence), and leucocytes (blue) (Figure 1).

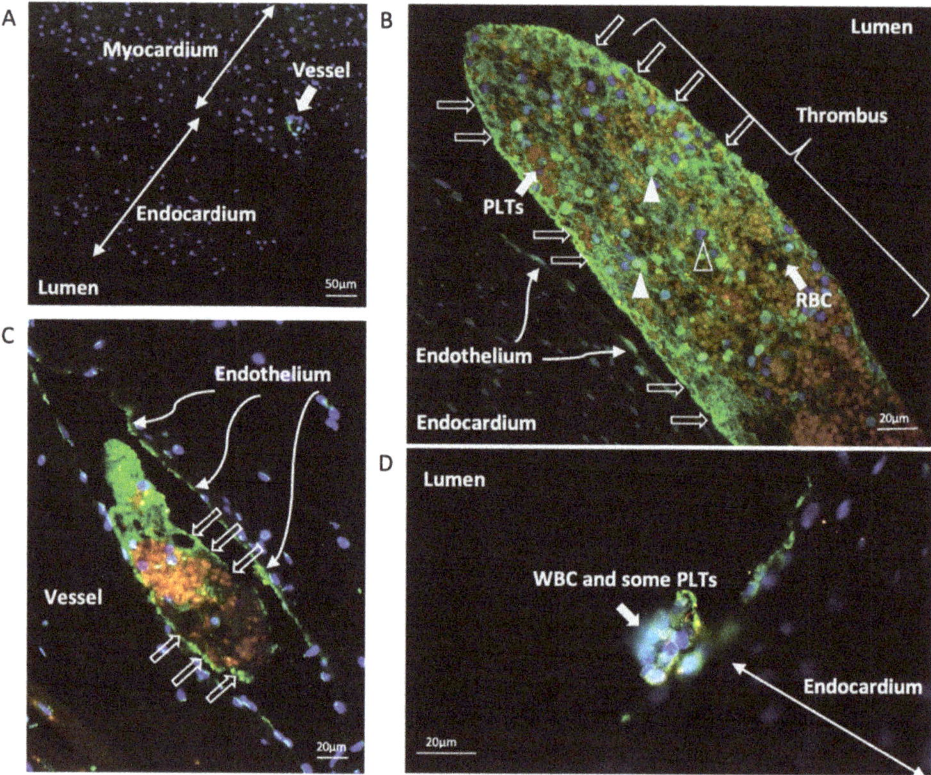

Figure 1. Localisation of vWF in the LA samples. (**A**) vWF localised to the endothelium of a vessel (wide arrow) in a control cat where there was minimal immunostaining of vWF at the left atrial endocardium. 100× magnification. (**B**) In this microthrombus from a cardiomyopathic cat with clinical signs, vWF appeared to form a scaffold outside (open arrows) and within the microthrombus. Red blood cells (RBC) were relatively fluorescence-lucent (thick arrow). vWF also localised to the leucocytes, (closed arrow heads) and platelets (PLTs) within the microthrombus. Some leucocytes did not immunostain for vWF (open arrow heads). Endocardial endothelial cells (thin arrows) also expressed vWF. (**C**) vWF localised to the endothelium of a vessel in a cat with cardiomyopathy and clinical signs. The intravascular microthrombus had vWF (green) encompassing the thrombus (open arrows) and contacting the vascular endothelium. Platelets and vWF co-localised in the centre of the microthrombus (orange, or yellow). (**D**) vWF (green), leucocytes, or WBC (blue) and some platelets (orange or yellow) attached on the endocardium from a cardiomyopathic cat with clinical signs. 400× magnification.

3.3. Quantification of Endocardial vWF Expression in Left Atrial Samples

The intensity of vWF fluorescence at the endocardium (Figure 2) was quantified using ImageJ for comparison between groups. To avoid quantifying the fluorescence signals from platelets where vWF can also be detected, the slides were double immunostained for vWF and the platelet marker integrin αIIb (Figure 3).

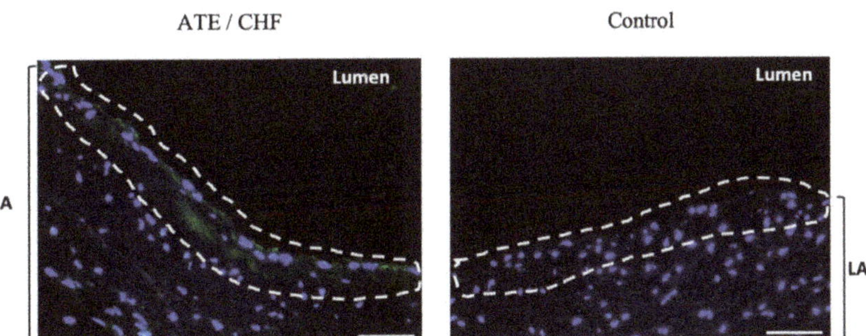

Figure 2. Localisation of vWF in the endocardium. Immunostaining of vWF in the endocardium. On the left, endocardial vWF (green) in a cat with cardiomyopathy and clinical signs. On the right, minimal endocardial immuostaining of vWF in a control cat. 400× magnification, Bar = 50 μm.

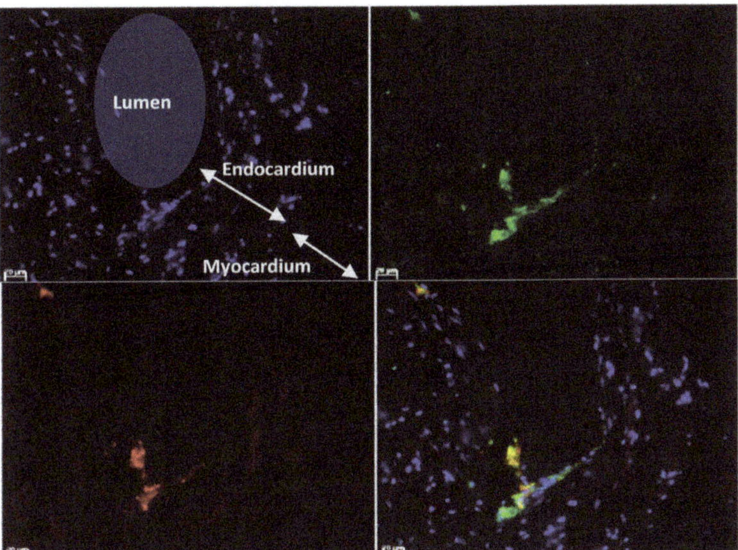

Figure 3. Double immunostaining for vWF and integrin αIIb. In addition to vWF, each slide was also stained for Integrin αIIb, a platelet marker, so the quantification of endocardial vWF would not include any platelets which are known to also express vWF. Images were displayed as split channels with the merged image on the bottom right. 400× magnification; bar = 20 μm; Blue—Nuclei; Green—vWF; Red—Integrin αIIb (CD41).

Representative images of vWF immunolabelling in different groups of cats are shown in Figure 4. Medians of the detected fluorescence intensity of endocardial vWF in the left atrial endocardium from the four groups from left to right as shown in Figure 5 were Control group: 30.8 (IQR 28.1–34.2), Subclinical 39.6 (IQR 31.8–59.6), CHF group: 46.0 (IQR 36.6–56.8), and ATE group: 44.7 (IQR 34.9–54.6). The fluorescence intensity was significantly higher in the ATE and CHF groups compared to the control group.

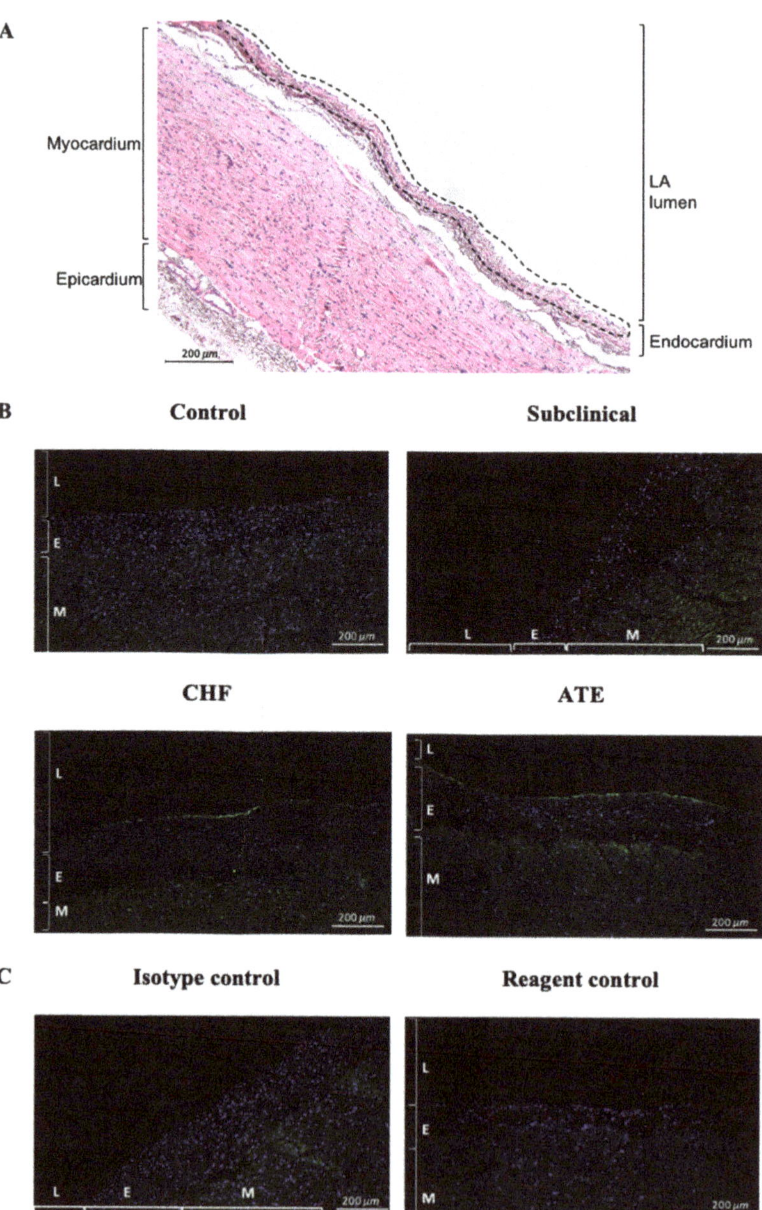

Figure 4. Comparison of the vWF immunostaining at the endocardium. (**A**) Image illustrating the LA anatomical structures and the region of interest where the intensity of immunofluorescence was quantified (dotted overlay). (**B**) Representative images showed the variation in immunolabelling of vWF (green) at the endocardium in the different groups of cats. (**C**) Isotype control immunostaining was performed using rabbit IgG. Reagent control was performed with the primary antibody omitted. $100\times$ magnification; Bar = 100 μm; LA (Left atrium), L (Lumen), E (Endocardium), M (Myocardium).

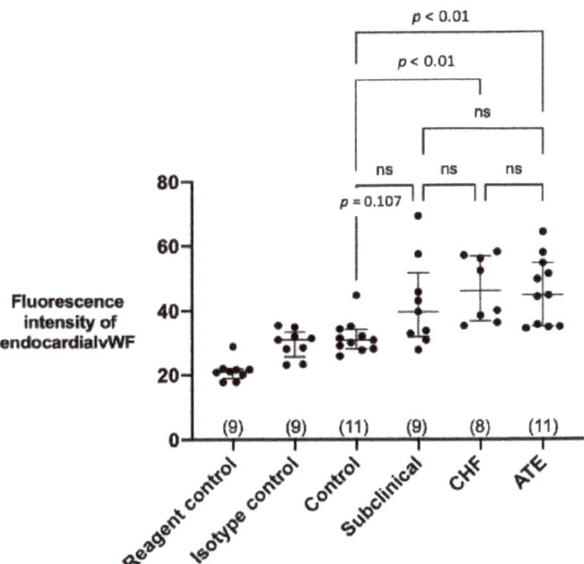

Figure 5. Quantification of the endocardial fluorescence from vWF: Cats with cardiomyopathy and clinical signs showed higher expression of vWF at the endocardium. The fluorescence signals detected in the ATE and CHF group were significantly higher compared to that of control group. No difference was detected between the rest of the groups. "ns" denotes non-significant. The number of cats in each group was shown in brackets. Data were analysed using Kruskal-Wallis test with Dunn's multiple comparison test. Bars represent median and quartiles.

3.4. Characterisation of vWF in Thrombi Obtained from LA and LAA

Thrombi were found in the LA or LAA in three cats (CHF 8, ATE 9, and ATE 12 in Tables S3 and S4) at postmortem. The necropsy images of the thrombi and the hearts can be found in Figure S1. The microscopic images of the thrombi, two retrieved from the LA and one remaining in situ in the LAA, are shown in Figure 6. These thrombi were immunostained for vWF (green) and platelets (reddish orange) and were counterstained with the nucleus stain (DAPI). WBC can be identified by the blue round-shaped nucleus stained by DAPI. RBC can be easily identified by their autofluorescence in the reagent controls (Figure 6, autofluorescence column) [41]. The autofluorescence of RBC was markedly weaker compared to the fluorophores bound to the antibodies used for labelling vWF, platelets and nuclei and thus appeared relatively dark in Figure 1, Figure 6, and Figure 7. The composition of each thrombi was very different in terms of relative proportions of the main components and how different components were organised. An enlarged image of the upper right quadrant of the LA thrombus from CHF 8 (Figure 6) is used as an example to show the various patterns of organisation (Figure 7A). Different patterns of vWF expression were also observed and enlarged images of these different patterns of expression are described in Figure 7B.

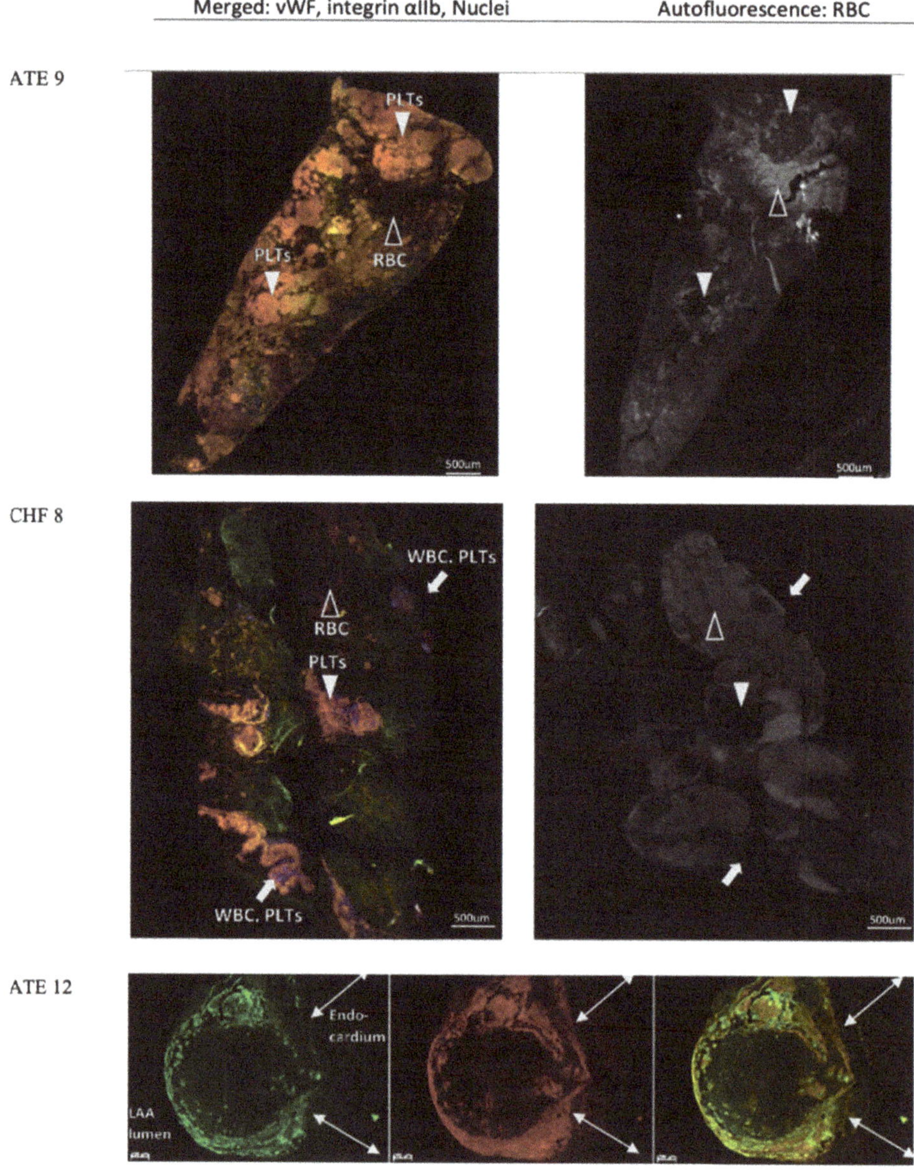

Figure 6. Microscopic images of thrombi immunostained for vWF and integrin αIIb: Images from Cats ATE 9, CHF 8: Images on the left shows the merged channels of vWF (green), platelet marker integrin αIIb (reddish orange) and nuclei (blue). In the merged channel, bright orange colour (closed arrow heads) represented clumps of platelets and vWF, while the RBC appeared dark and non-fluorescent (open arrow heads) compared to the immunolabelled vWF and platelets. WBCs were identified based on the blue colour of their nuclei and could be observed with platelets in clumps (arrows). Images were collated to show the whole thrombus. Images from Cat ATE 12: Split channels from left to right showed vWF in green, platelet marker integrin αIIb in red and the merged of the two channels. This thrombus remained in situ in the LAA. vWF and platelets encompassed RBC and connected to the endocardium. 50× magnification; bar = 200 μm. RBC, red blood cells; PLT, platelets; WBC, white blood cells; LAA, left atrial appendage.

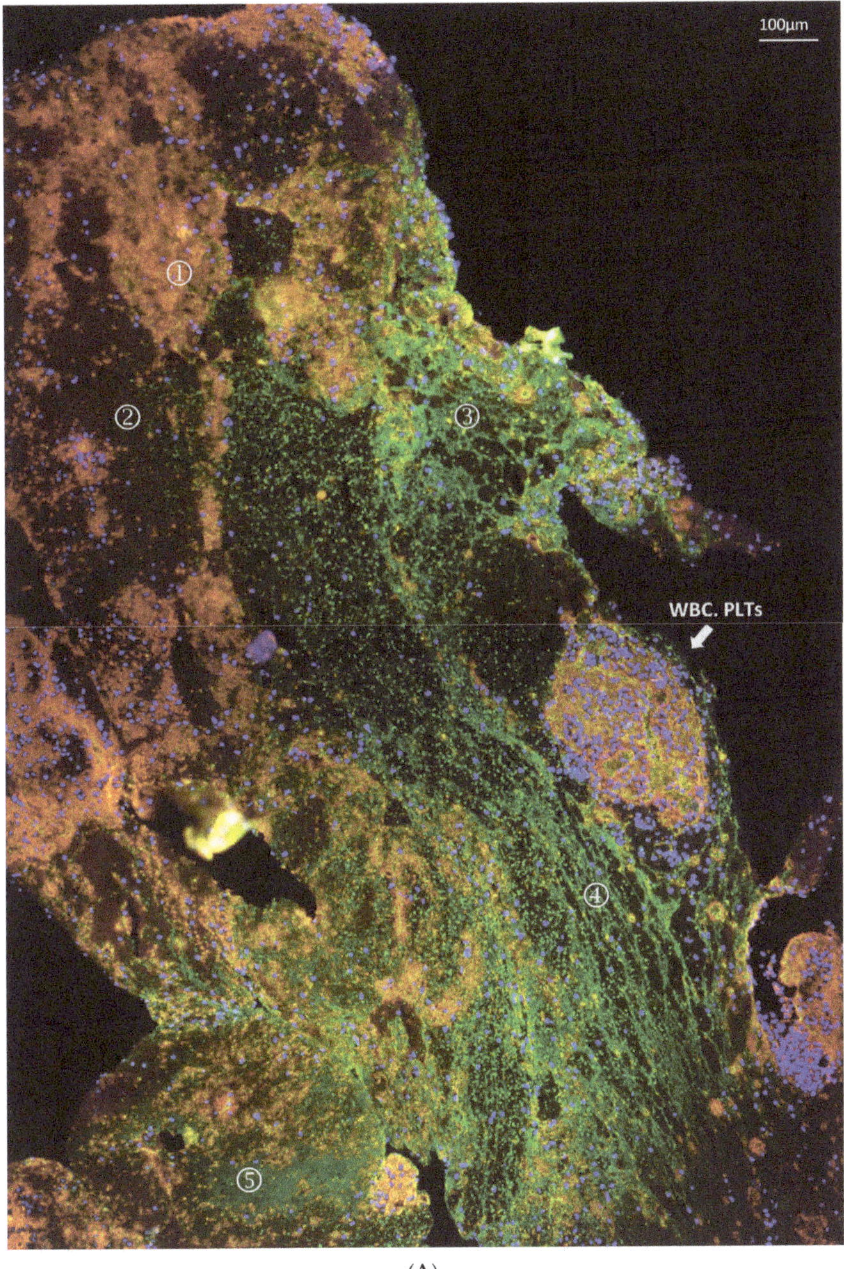

(A)

Figure 7. *Cont.*

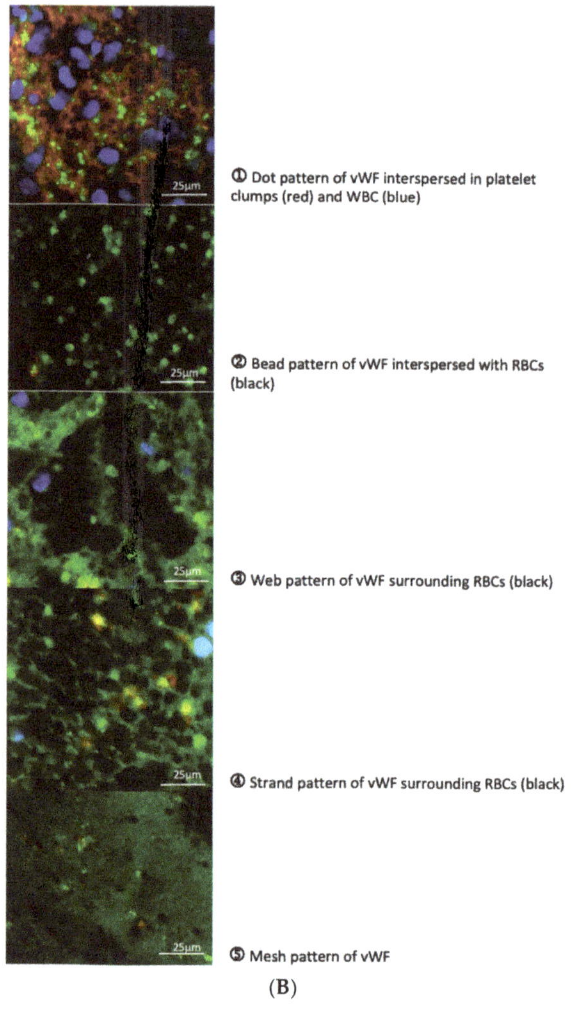

(**B**)

Figure 7. Enlarged image of the LA thrombus from Cat CHF 8. The clump of WBCs and platelets denoted by a thick white arrow in (**A**) is the same area of the thrombus from CAT CHF 8 shown in the middle left hand image of Figure 6. Different patterns of vWF staining and the variety of blood cell types in different areas of the thrombus are shown in (**B**) 1–5: 1—dots interspersed in clumps of platelets; 2—beads interspersed within RBC; 3—web surrounding RBC; 4—strands surrounding RBC; 5—mesh with minimal platelets, RBC, or WBC.

4. Discussion

Cats with underlying myocardial disease are known to be at a greater risk of thromboembolic events which frequently results in euthanasia on humane grounds [6,7]. Given the high prevalence of cardiomyopathy in cats, which for HCM is estimated at up to 17% in outbred animals and even greater in certain pedigree breeds [33], it is important to obtain a greater understanding of the left atrial remodelling process that predisposes to thrombus formation.

This is the first study to quantify and characterise vWF in the LA endocardium and LA thrombi from cats with cardiomyopathy at different stages of their disease process. LA samples from cardiomyopathic cats with LAE and clinical signs (CHF and ATE), but not those from subclinical cardiomyopathic cats had increased endocardial vWF protein compared to control cats (Figure 5), which suggests an association of endocardial vWF elevation with an advanced stage of cardiomyopathy. The immunohistological evaluation of the thrombi revealed that vWF was extensively involved in thrombus formation and organisation of both microthrombi and the three macrothrombi, which were visible on gross pathology within the LA(A) in cats CHF 8, ATE 9, and ATE 12 shown in Figure S1.

The finding that vWF localised to some vessels in the LA samples is consistent with previous report of differential expression of endothelial vWF in mice where vWF expression varied not just in different organs but even in the vessels in the same vascular tree [42]. The minimal detection of vWF in the endocardium in the control cats is similar to that in a human report where no or only minimal focal immunostaining for vWF was present in the LAA endocardium from patients without cardiac abnormalities [19,43]. It is also striking that there were no female cats in our ATE group, which is consistent with previous reports showing a male predisposition to ATE and to an increase hazard of an ATE associated death. However, the possibility of this finding being due to chance cannot be ruled out [7,44].

Importantly, our finding in cats with advanced myocardial disease are similar to two human studies where patients with hemodynamical disturbance in the LA secondary to various cardiac causes showed elevated endocardial vWF compared to the non-cardiac patients using immunohistological microscopy [19,20]. In the first of these studies, a significant increase in the expression of vWF in the endocardium of atrial appendages was identified in human patients with a variety of congenital and acquired heart diseases irrespective of the presence of atrial fibrillation. Furthermore, increased vWF expression correlated with the degree of platelet adhesion and thrombus formation [19]. In the second study involving patients with valvular and non-valvular atrial fibrillation, expression of vWF protein in the endocardium correlated with the degree of structural remodelling in the atrial wall. In addition, endocardial vWF appeared important for platelet adhesion/aggregation on the endocardium and intra-atrial thrombi formation suggesting that increased endocardial vWF may contribute to LA thrombogenesis [20]. It is recognised that inflammation per se can cause thrombosis via a VWF-mediated mechanism by elevating the level of VWF, enhancing the reactivity of VWF, and modulating the levels and activities of regulatory molecules such as the metalloprotease ADAMTS13 [45]. Heart failure is an established cause of systemic inflammation, and the increased expression of vWF in our cohort of cats with clinical signs may at least in part be related to this pro-inflammatory milieu, however the relationship between circulating concentrations of vWF and heart failure remains unclear in human studies [46–48] and that between endocardial vWF and inflammation is not examined.

The immunostaining of the micro- and macrothrombi supports a pivotal role for vWF in thrombus formation. Thrombosis is a multifactorial process which usually involves an abnormality in one or more of the following factors: endothelial damage and dysfunction, disturbed blood flow, and hypercoagulability. Thrombus formation begins when platelets adhere to the exposed subendothelial matrix by means of vWF that was previously deposited on the sub-endothelium by endothelial cells. Once adhered, the stimulated platelets then secrete more mediators such as ADP and thromboxane A2 that further activate additional platelets leading to propagation of the thrombus [49,50].

Traditionally, two categories of thrombus have been described based on their gross appearance and composition. A red thrombus that forms under low shear stress and is primarily composed of erythrocytes and fibrin, and a white thrombus that forms under high shear stress and contains a large number of platelets and few erythrocytes. The former mostly comprise venous thrombi where stasis of blood is evident and endothelial injury is not an absolute requirement for thrombus formation. Conversely, white thrombi are usually confined to the arterial system and form after the severe endothelial disruption such as plaque erosion or rupture as part of the atherosclerosis process [51]. In the present study, platelets were found to be an important component in the thrombi despite being formed in the environment of slow blood flow and low shear stress found in the enlarged

LA. Different patterns of vWF expression in the thrombi were seen in or around particular cellular components forming different structural patterns around the same cell type, such us bead-like or web-like patterns around erythrocytes. vWF is known to fold or unfold under different conditions of shear stress and can therefore display various properties and functions as a result of exposing certain binding sites for adhesion or uncovering its cleavage site [52,53]. These various structural patterns formed by the thrombi might imply that the local environment within the LA where the thrombus developed altered over time, affecting vWF function and thereby the composition of the thrombus over its development. More research is needed to investigate how and why the feline cardiogenic thrombi showed components of both a traditional red and white thrombus.

There are a number of limitations to the study. First, the origin of the detected endocardial vWF protein in the LA samples remains undetermined. We cannot rule out the possibility that the LA endocardial vWF we detected was originally secreted by platelets or endothelia elsewhere in the body and delivered via the circulation. Second, although all the control cats had structurally normal hearts, a number did have other clinical potentially inflammatory conditions that may have affected endothelial vWF expression. However, most systemic inflammatory diseases are associated with increased level of vWF, which is contrary to the finding of low endocardial vWF in the control cats in our study. Third, in the subclinical group, four out of the nine cats had mild LAE. However, when we further sub-divided these cats into (1) subclinical with LAE, and (2) subclinical without LAE, there was no significant difference in endocardial fluorescence intensity between these subgroups and the controls. Fourth, to better evaluate the structure of the thrombi, sliced-though slides with co-immunostaining for fibrin would be helpful to gain greater understanding of thrombus organisation. However, this was beyond the scope of the study and not performed.

5. Conclusions

These results provide evidence for increased endocardial vWF in cats with advanced cardiomyopathy and support a potential role for endocardial vWF as a pro-thrombotic substrate.

Supplementary Materials: The following are available online at https://www.mdpi.com/article/10.3390/ani11051240/s1, Figure S1: Gross appearance of the thrombi. Table S1: Pathology classification criteria for feline cardiomyopathies, Table S2: Echocardiographic classification criteria for feline cardiomyopathies, Table S3: Clinical presentation and echocardiography summary, Table S4: Signalment, echocardiographic parameters, and histopathology diagnosis, Table S5: Measurement of echocardiographic variables. References [33,36,38,39] are also cited in the supplementary materials.

Author Contributions: W.-C.C. contributed to the design of the study and performed the majority of the laboratory work. She also analysed the results and wrote the manuscript. L.W. performed the gross and histopathology on the specimens and contributed to writing the manuscript. T.A.K. contributed to the design of the study and some of the laboratory work as well as contributing to writing the manuscript. M.D. performed the gross and histopathology on the specimens and contributed to writing the manuscript. S.L.P. contributed to the design the study and to writing the manuscript and gave editorial advice. V.L.F. contributed to the design of the study and to writing the manuscript and gave editorial advice. D.J.C. was involved in study design and day to day supervision of the study as well as a primary contributor to the manuscript and its subsequent edits. All authors have read and agreed to the published version of the manuscript.

Funding: The author Wan-Ching Cheng received funding from Taiwan Ministry of Education.

Institutional Review Board Statement: The study was and approved by the Ethics Committee of the Royal Veterinary College—approval number URN 2014 1301.

Data Availability Statement: The data presented in this study are available on request from the corresponding author. The data are not publicly available due to privacy issues.

Acknowledgments: We would like to acknowledge Yu-Mei Ruby Chang and Elisabeth Ehler for the assistance and advice with the statistical analysis and immunostaining, respectively.

Conflicts of Interest: The authors declare no conflict of interest.

References

1. Smith, S.A.; Tobias, A.H.; Jacob, K.A.; Fine, D.M.; Grumbles, P.L. Arterial Thromboembolism in Cats: Acute Crisis in 127 Cases (1992–2001) and Long-Term Management with Low-Dose Aspirin in 24 Cases. *J. Vet. Intern. Med.* **2003**, *17*, 73–83. [CrossRef]
2. Laste, N.J.; Harpster, N.K. A retrospective study of 100 cases of feline distal aortic thromboembolism: 1977–1993. *J. Am. Anim. Hosp. Assoc.* **1995**, *31*, 492–500. [CrossRef] [PubMed]
3. Hogan, D.F. Feline Cardiogenic Arterial Thromboembolism. *Vet. Clin. N. Am. Small Anim. Pr.* **2017**, *47*, 1065–1082. [CrossRef] [PubMed]
4. Fox, P.R.; Keene, B.W.; Lamb, K.; Schober, K.A.; Chetboul, V.; Fuentes, V.L.; Wess, G.; Payne, J.R.; Hogan, D.F.; Motsinger-Reif, A.; et al. International collaborative study to assess cardiovascular risk and evaluate long-term health in cats with preclinical hypertrophic cardiomyopathy and apparently healthy cats: The REVEAL Study. *J. Vet. Intern. Med.* **2018**, *32*, 930–943. [CrossRef] [PubMed]
5. Payne, J.; Fuentes, V.L.; Boswood, A.; Connolly, D.; Koffas, H.; Brodbelt, D. Population characteristics and survival in 127 referred cats with hypertrophic cardiomyopathy (1997 to 2005). *J. Small Anim. Pr.* **2010**, *51*, 540–547. [CrossRef]
6. Fox, P.R.; Liu, S.-K.; Maron, B.J. Echocardiographic Assessment of Spontaneously Occurring Feline Hypertrophic Cardiomyopathy: An animal model of human disease. *Circulation* **1995**, *92*, 2645–2651. [CrossRef]
7. Borgeat, K.; Wright, J.; Garrod, O.; Payne, J.; Fuentes, V.L. Arterial Thromboembolism in 250 Cats in General Practice: 2004–2012. *J. Veter Intern. Med.* **2013**, *28*, 102–108. [CrossRef]
8. Bagot, C.N.; Arya, R. Virchow and his triad: A question of attribution. *Br. J. Haematol.* **2008**, *143*, 180–190. [CrossRef]
9. Schober, K.E.; Maerz, I. Assessment of left atrial appendage flow velocity and its relation to spontaneous echocardiographic contrast in 89 cats with myocardial disease. *J. Vet. Intern. Med.* **2006**, *20*, 120–130. [CrossRef]
10. Stokol, T.; Brooks, M.; Rush, J.; Rishniw, M.; Erb, H.; Rozanski, E.; Kraus, M.; Gelzer, A. Hypercoagulability in Cats with Cardiomyopathy. *J. Veter Intern. Med.* **2008**, *22*, 546–552. [CrossRef]
11. Liu, S.K.; Maron, B.J.; Tilley, L.P. Feline hypertrophic cardiomyopathy: Gross anatomic and quantitative histologic features. *Am. J. Pathol.* **1981**, *102*, 388–395.
12. McMichael, M.; Freeman, L.; Selhub, J.; Rozanski, E.; Brown, D.; Nadeau; Rush, J. Plasma Homocysteine, B Vitamins, and Amino Acid Concentrations in Cats with Cardiomyopathy and Arterial Thromboembolism. *J. Vet. Intern. Med.* **2000**, *14*, 507–512. [CrossRef]
13. Ruggeri, Z.M. The role of von Willebrand factor in thrombus formation. *Thromb. Res.* **2007**, *120*, S5–S9. [CrossRef] [PubMed]
14. Zhou, Y.-F.; Eng, E.T.; Zhu, J.; Lu, C.; Walz, T.; Springer, T.A. Sequence and structure relationships within von Willebrand factor. *Blood* **2012**, *120*, 449–458. [CrossRef]
15. Reinders, J.H.; De Groot, P.G.; Sixma, J.J.; Van Mourik, J.A. Storage and Secretion of von Willebrand Factor by Endothelial Cells. *Pathophysiol. Haemost. Thromb.* **1988**, *18*, 246–261. [CrossRef] [PubMed]
16. Miura, S.; Li, C.Q.; Cao, Z.; Wang, H.; Wardell, M.R.; Sadler, J. Interaction of von Willebrand Factor Domain A1 with Platelet Glycoprotein Ibα-(1–289). *J. Biol. Chem.* **2000**, *275*, 7539–7546. [CrossRef]
17. Cosemans, J.M.E.M.; Schols, S.E.M.; Stefanini, L.; De Witt, S.; Feijge, M.A.H.; Hamulyák, K.; Deckmyn, H.; Bergmeier, W.; Heemskerk, J.W.M. Key role of glycoprotein Ib/V/IX and von Willebrand factor in platelet activation-dependent fibrin formation at low shear flow. *Blood* **2011**, *117*, 651–660. [CrossRef] [PubMed]
18. Lip, G.Y. von Willebrand factor: A marker of endothelial dysfunction in vascular disorders? *Cardiovasc. Res.* **1997**, *34*, 255–265. [CrossRef]
19. Fukuchi, M.; Watanabe, J.; Kumagai, K.; Katori, Y.; Baba, S.; Fukuda, K.; Yagi, T.; Iguchi, A.; Yokoyama, H.; Miura, M.; et al. Increased von Willebrand factor in the endocardium as a local predisposing factor for thrombogenesis in overloaded human atrial appendage. *J. Am. Coll. Cardiol.* **2001**, *37*, 1436–1442. [CrossRef]
20. Kumagai, K.; Fukuchi, M.; Ohta, J.; Baba, S.; Oda, K.; Akimoto, H.; Kagaya, Y.; Watanabe, J.; Tabayashi, K.; Shirato, K. Expression of the von Willebrand Factor in Atrial Endocardium is Increased in Atrial Fibrillation Depending on the Extent of Structural Remodeling. *Circ. J.* **2004**, *68*, 321–327. [CrossRef]
21. Ammash, N.; Konik, E.A.; McBane, R.D.; Chen, D.; Tange, J.I.; Grill, D.E.; Herges, R.M.; McLeod, T.G.; Friedman, P.A.; Wysokinski, W.E. Left Atrial Blood Stasis and Von Willebrand Factor–ADAMTS13 Homeostasis in Atrial Fibrillation. *Arter. Thromb. Vasc. Biol.* **2011**, *31*, 2760–2766. [CrossRef] [PubMed]
22. Watson, T.; Shantsila, E.; Lip, G.Y. Mechanisms of thrombogenesis in atrial fibrillation: Virchow's triad revisited. *Lancet* **2009**, *373*, 155–166. [CrossRef]
23. Ding, W.Y.; Gupta, D.; Lip, G.Y.H. Atrial fibrillation and the prothrombotic state: Revisiting Virchow's triad in 2020. *Hear.* **2020**, *106*, 1463–1468. [CrossRef] [PubMed]
24. Spiel, A.O.; Gilbert, J.C.; Jilma, B. Von Willebrand Factor in Cardiovascular Disease: Focus on acute coronary syndromes. *Circulation* **2008**, *117*, 1449–1459. [CrossRef]

25. Sanders, Y.V.; Eikenboom, J.; De Wee, E.M.; Van Der Bom, J.G.; Cnossen, M.H.; Degenaar-Dujardin, M.E.L.; Fijnvandraat, K.; Kamphuisen, P.W.; Gorkom, B.A.P.L.-V.; Meijer, K.; et al. Reduced prevalence of arterial thrombosis in von Willebrand disease. *J. Thromb. Haemost.* **2013**, *11*, 845–854. [CrossRef] [PubMed]
26. Brill, A.; Fuchs, T.A.; Chauhan, A.K.; Yang, J.J.; De Meyer, S.F.; Koellnberger, M.; Wakefield, T.W.; Laemmle, B.; Massberg, S.; Wagner, D.D. von Willebrand factor–mediated platelet adhesion is critical for deep vein thrombosis in mouse models. *Blood* **2011**, *117*, 1400–1407. [CrossRef] [PubMed]
27. Carvalho, F.A.; Connell, S.; Miltenberger-Miltenyi, G.; Pereira, S.V.; Tavares, A.; Ariëns, R.A.S.; Santos, N.C. Atomic Force Microscopy-Based Molecular Recognition of a Fibrinogen Receptor on Human Erythrocytes. *ACS Nano* **2010**, *4*, 4609–4620. [CrossRef] [PubMed]
28. De Oliveira, S.; de Almeida, V.V.; Calado, A.; Rosário, H.; Saldanha, C. Integrin-associated protein (CD47) is a putative mediator for soluble fibrinogen interaction with human red blood cells membrane. *Biochim. et Biophys. Acta (BBA) Biomembr.* **2012**, *1818*, 481–490. [CrossRef]
29. Aleman, M.M.; Walton, B.L.; Byrnes, J.R.; Wolberg, A.S. Fibrinogen and red blood cells in venous thrombosis. *Thromb. Res.* **2014**, *133*, S38–S40. [CrossRef]
30. Walton, B.L.; Byrnes, J.R.; Wolberg, A.S. Fibrinogen, red blood cells, and factor XIII in venous thrombosis. *J. Thromb. Haemost.* **2015**, *13*, S208–S215. [CrossRef] [PubMed]
31. Smeets, M.W.J.; Mourik, M.J.; Niessen, H.W.M.; Hordijk, P.L. Stasis Promotes Erythrocyte Adhesion to von Willebrand Factor. *Arter. Thromb. Vasc. Biol.* **2017**, *37*, 1618–1627. [CrossRef] [PubMed]
32. Wilkie, L.; Smith, K.; Fuentes, V.L. Cardiac pathology findings in 252 cats presented for necropsy; a comparison of cats with unexpected death versus other deaths. *J. Veter Cardiol.* **2015**, *17*, S329–S340. [CrossRef]
33. Wilkie, L.J. Phenotypic progression and cardiac remodelling in feline cardiomyopathies. Ph.D. Thesis, Royal Veterinary College, University of London, Hatfield, UK, 2017.
34. Payne, J.R.; Brodbelt, D.C.; Fuentes, V.L. Cardiomyopathy prevalence in 780 apparently healthy cats in rehoming centres (the CatScan study). *J. Veter Cardiol.* **2015**, *17*, S244–S257. [CrossRef]
35. Fuentes, V.L.; Abbott, J.; Chetboul, V.; Côté, E.; Fox, P.R.; Häggström, J.; Kittleson, M.D.; Schober, K.; Stern, J.A. ACVIM consensus statement guidelines for the classification, diagnosis, and management of cardiomyopathies in cats. *J. Veter Intern. Med.* **2020**, *34*, 1062–1077. [CrossRef] [PubMed]
36. Côté, E.; Macdonald, K.A.; Meurs, K.M.; Sleeper, M.M. *Feline Cardiology*, 1st ed.; John Wiley & Sons, Inc.: West Sussex, UK, 2011; ISBN 9781118785782.
37. Seo, J.; Payne, J.R.; Matos, J.N.; Fong, W.W.; Connolly, D.J.; Fuentes, V.L. Biomarker changes with systolic anterior motion of the mitral valve in cats with hypertrophic cardiomyopathy. *J. Veter Intern. Med.* **2020**, *34*, 1718–1727. [CrossRef] [PubMed]
38. Hansson, K.; Haggstrom, J.; Kvart, C.; Lord, P. Left atrial to aortic root indices using two-dimensional and m-mode echocardiography in cavalier king charles spaniels with and without left atrial enlargement. *Vet. Radiol. Ultrasound* **2002**, *43*, 568–575. [CrossRef] [PubMed]
39. März, I.; Wilkie, L.J.; Harrington, N.; Payne, J.R.; Muzzi, R.A.L.; Häggström, J.; Smith, K.; Fuentes, V.L. Familial cardiomyopathy in Norwegian Forest cats. *J. Feline Med. Surg.* **2014**, *17*, 681–691. [CrossRef]
40. Fuentes, V.L. Arterial Thromboembolism: Risks, realities and a rational first-line approach. *J. Feline Med. Surg.* **2012**, *14*, 459–470. [CrossRef]
41. Whittington, N.C.; Wray, S. Suppression of Red Blood Cell Autofluorescence for Immunocytochemistry on Fixed Embryonic Mouse Tissue. *Curr. Protoc. Neurosci.* **2017**, *81*, 2.28.1–2.28.12. [CrossRef]
42. Yamamoto, K.; De Waard, V.; Fearns, C.; Loskutoff, D.J. Tissue Distribution and Regulation of Murine von Willebrand Factor Gene Expression In Vivo. *Blood* **1998**, *92*, 2791–2801. [CrossRef]
43. Nakamura, Y.; Nakamura, K.; Fukushima-Kusano, K.; Ohta, K.; Matsubara, H.; Hamuro, T.; Yutani, C.; Ohe, T. Tissue factor expression in atrial endothelia associated with nonvalvular atrial fibrillation: Possible involvement in intracardiac thrombogenesis. *Thromb. Res.* **2003**, *111*, 137–142. [CrossRef]
44. Payne, J.; Borgeat, K.; Brodbelt, D.; Connolly, D.; Fuentes, V.L. Risk factors associated with sudden death vs. congestive heart failure or arterial thromboembolism in cats with hypertrophic cardiomyopathy. *J. Veter Cardiol.* **2015**, *17*, S318–S328. [CrossRef] [PubMed]
45. Chen, J.; Chung, D.W. Inflammation, von Willebrand factor, and ADAMTS13. *Blood* **2018**, *132*, 141–147. [CrossRef] [PubMed]
46. Murphy, S.P.; Kakkar, R.; McCarthy, C.P.; Januzzi, J.L. Inflammation in Heart Failure. *J. Am. Coll. Cardiol.* **2020**, *75*, 1324–1340. [CrossRef] [PubMed]
47. Wannamethee, S.G.; Whincup, P.H.; Papacosta, O.; Lennon, L.; Lowe, G.D. Associations between blood coagulation markers, NT-proBNP and risk of incident heart failure in older men: The British Regional Heart Study: JACC State-of-the-Art Review. *Int. J. Cardiol.* **2017**, *230*, 567–571. [CrossRef]
48. Nanchen, D.; Stott, D.J.; Gussekloo, J.; Mooijaart, S.P.; Westendorp, R.G.; Jukema, J.W.; Macfarlane, P.W.; Cornuz, J.; Rodondi, N.; Buckley, B.M.; et al. Resting heart rate and incident heart failure and cardiovascular mortality in older adults: Role of inflammation and endothelial dysfunction: The PROSPER study. *Eur. J. Hear. Fail.* **2013**, *15*, 581–588. [CrossRef]
49. Furie, B.; Furie, B.C. Thrombus formation in vivo. *J. Clin. Investig.* **2005**, *115*, 3355–3362. [CrossRef]
50. Furie, B.; Furie, B.C. Mechanisms of Thrombus Formation. *New Engl. J. Med.* **2008**, *359*, 938–949. [CrossRef]

51. Tan, K.T.; Lip, G.Y.H. Red vs White Thrombi: Treating the Right Clot Is Crucial. *Arch. Intern. Med.* **2003**, *163*, 2534–2535. [CrossRef]
52. Tsai, H.-M. Shear Stress and von Willebrand Factor in Health and Disease. *Semin. Thromb. Hemost.* **2003**, *29*, 479–488. [CrossRef]
53. Tsai, H.-M. von Willebrand Factor, Shear Stress, and ADAMTS13 in Hemostasis and Thrombosis. *ASAIO J.* **2012**, *58*, 163–169. [CrossRef] [PubMed]

Article

Immunohistochemical Expression of Neurokinin-A and Interleukin-8 in the Bronchial Epithelium of Horses with Severe Equine Asthma Syndrome during Asymptomatic, Exacerbation, and Remission Phase

Maria Morini [1,*], Angelo Peli [1], Riccardo Rinnovati [1], Giuseppe Magazzù [2], Noemi Romagnoli [1], Alessandro Spadari [1] and Marco Pietra [1]

[1] Department of Veterinary Medical Sciences, University of Bologna, 40064 Bologna, Italy; angelo.peli@unibo.it (A.P.); riccardo.rinnovati2@unibo.it (R.R.); noemi.romagnoli@unibo.it (N.R.); alessandro.spadari@unibo.it (A.S.); marco.pietra@unibo.it (M.P.)

[2] DVM, Vet Practitioner, 40024 Castel San Pietro Terme, 40064 Bologna, Italy; magazzu.consultant@gmail.com

[*] Correspondence: maria.morini@unibo.it; Tel.: +39-051-209-7970

Citation: Morini, M.; Peli, A.; Rinnovati, R.; Magazzù, G.; Romagnoli, N.; Spadari, A.; Pietra, M. Immunohistochemical Expression of Neurokinin-A and Interleukin-8 in the Bronchial Epithelium of Horses with Severe Equine Asthma Syndrome during Asymptomatic, Exacerbation, and Remission Phase. *Animals* **2021**, *11*, 1376. https://doi.org/10.3390/ani11051376

Academic Editors: Alejandro Suárez-Bonnet and Gustavo A. Ramírez Rivero

Received: 7 April 2021
Accepted: 6 May 2021
Published: 12 May 2021

Publisher's Note: MDPI stays neutral with regard to jurisdictional claims in published maps and institutional affiliations.

Simple Summary: Severe equine asthma (EA) syndrome, formerly termed Recurrent Airway Obstruction (RAO) or heaves, is one of the most common respiratory diseases in adult horses and a frequent cause of poor equine performance. The affected animals may show periods of clinical remission followed by periods of exacerbation over months to years. Therefore, the aim of the present study was to investigate the histological features, and the Neurokinin-A (NKA) and Interleukin-8 (IL-8) immunoreactivity on bronchoscopic biopsies in horses, obtained during different phases of the disease (asymptomatic, exacerbation and remission). Histological samples of EA-affected horses appeared significantly different from those of non-EA-affected horses (control group) throughout the experimental phase, from inclusion to exacerbation and remission, and intensity of NKA immunopositivity of horses with severe EA was significantly higher than that of control horses in late exacerbation and in remission phase. No significant difference between horses with severe EA in each phase and control horses was noticed for IL-8 immunoreactivity. Moreover, no influence of bronchial sampling position on histological and immunohistochemistry results was found, and it suggests that bronchial structural and functional modification during severe equine asthma tends to be distributed homogeneously throughout the respiratory tree.

Abstract: Severe equine asthma (EA) syndrome is a chronic obstructive disease characterized by exaggerated contraction, inflammation, and structural alteration of the airways in adult horses, when exposed to airborne molds and particulate material. However, little is known about the relationship between the degree and type of inflammation on one hand, and the severity of the disease and the response to treatment on the other. Furthermore, to date, very few studies evaluate the diagnostic value of histology and immunohistochemical features of endoscopic biopsies on subjects with severe equine asthma. To investigate the expression of two inflammatory markers (NKA and IL-8) before, during, and after the exacerbation of severe EA, a histological and immunohistochemical study was carried out on a series of biopsy samples collected by bronchoscopy from six EA-affected horses subjected to process exacerbation through environmental stimuli and then to pharmacological treatment. The application of a histological biopsy scoring system revealed a significant difference between control cases and the EA-affected horses in all experimental phases (asymptomatic, early exacerbation phase, late exacerbation phase, and remission phase). For immunohistochemistry (IHC), only the intensity of NKA positivity increases significantly between control horses and the EA horses at late exacerbation and remission phases. In EA-affected horses, a difference was detected by comparing histology between asymptomatic and remission phase, meanwhile, NKA and IL-8 showed no differences between the experimental phases. Based on these results we can assert that: (1) The endoscopic biopsies generate reliable and homogeneous samples in the entire bronchial tree; (2) the clinical improvement associated with treatment is characterized by a significant worsening of the histological findings; and (3) the NKA immunopositivity seems to increase significantly rather than decrease, as one would have expected, after pharmacological treatment. Further studies are necessary both to implement the number of samples and to use other markers of inflammation to characterize the potential role of cytokines in the diagnosis and therapeutic approach of severe equine asthma.

Keywords: respiratory tract; horse; severe equine asthma; immunohistochemistry; IL-8; NKA

1. Introduction

Disorders of the respiratory system, particularly the lower airways, are the most frequently diagnosed conditions in sport horses evaluated for poor performance [1]. The terminology used to describe equine chronic noninfectious small airway disease has further evolved in the last few years with the term "equine asthma" (EA), as new features (functional, anatomical and pathobiological) of this condition are emerging [2,3]. EA is now being recommended to describe horses with chronic respiratory signs ranging in severity from mild to severe: they were previously referred as inflammatory airway disease (IAD) or recurrent airway obstruction (RAO), respectively [3]. RAO is then considered as one of the main features of horses with a severe form of equine asthma (also defined as severe equine asthma syndrome) [2–6], which resembles human asthma in many aspects [3,4].

Horses affected by severe equine asthma syndrome show labored breathing at rest following exposure to specific airborne agents, and reversible and reproducible airway obstruction related to the level of environmental exposures [4]. The exposure to hay and dust leading to heaves is rather a consequence of the human influence on the horses' natural environment. Molds and fungi are indeed common antigens in stables, suggesting that EA is a disease of "domestication". However, horses can develop a similar condition while at pasture, with grass pollen then being the likely triggering factor [7–9].

Clinical exacerbation occurs following exposure of susceptible horses to specific airborne agents, and results in a disease phenotype of varying severity, ranging from exercise intolerance to coughing and severe expiratory dyspnea. As the name implies, the disease is largely reversible, whereby avoidance of the inciting airborne agents results in significant disease remission over time. The cardinal clinical features of severe EA can be due to the underlying airway inflammatory response that underpins the particular functional and pathological features of the disease [10].

Since bronchoalveolar lavage (BAL) by use of fiber-optic endoscopy was first described in horses, cytological and microbiological evaluation of tracheal washes and BAL fluid have become the cornerstones in the diagnosis of respiratory disease alongside clinical and functional examinations. In exacerbation of severe equine asthma, the horses show dyspnea at rest, as revealed by a maximum intrapleural pressure >15 cm H_2O caused by bronchoconstriction, mucosal swelling and mucus accumulation [11], and inflammation of the small airways, in which neutrophils exceed 25% in bronchoalveolar lavage fluid (BALF) cytology [2]. The definitive diagnosis of severe equine asthma is usually based on BALF cytology with the presence of lower airway inflammation, characterized by total nucleated cell count with mild increased numbers of neutrophils [12–14] and lymphocytes, and by increased mast cell or eosinophilic counts [15,16]. Therefore, two different cytological phenotypes are recognized in severe equine asthma: a classical neutrophilic phenotype and a paucigranulocytic phenotype. They do not correspond to the histopathological finding observable in the peripheral airway [2] and suggesting a complex role for pulmonary neutrophils in equine asthma pathophysiology. Currently, little is known about the degree and type of BALF inflammation, during the different phases of the disease (asymptomatic, exacerbation and remission phase). Moreover, pathological features of the bronchial mucosa and their value in the diagnosis of severe equine asthma syndrome are poorly characterized. To date, histological findings collected by endoscopic biopsies do not allow differentiating between controls and subjects with severe EA in remission [17,18].

The immunologic background of severe equine asthma remains not fully clarified despite many studies on the pathogenesis [19,20]. However, a positive correlation exists between the intensity of airway hyperreactivity and the quantity of chemical mediators released locally in the lung [21]. Generally, airway inflammation involves activation of a pathogen-specific inflammatory cell, the modulation of gene transcription factors, and release of inflammatory mediators [16]. Commonly accepted airway inflammatory mediators involved in airway disease include histamine, bradykinin, prostaglandin, leukotrienes, platelet-activating factor, and endothelin-1 [20–22]. A tachykinin mediator with a physiological and pathological role in respiratory function is a neuropeptide called neurokinin A (NKA), which is involved in the processes of bronchoconstriction and neurogenic inflammation in asthmatic patients, with potential therapeutic implications using selective NKA receptor antagonists [20]. To the best of our knowledge, only two studies investigated the role of neurokinin-A (NKA) in the horse respiratory tract [20,23], and only one by immunohistochemical methods [20].

Neutrophilic bronchiolitis is one of the main lesions of asthma-affected horse's response to aeroallergens. Within the airways, neutrophils likely contribute to bronchoconstriction, mucus hypersecretion, and pulmonary remodeling by release of pro-inflammatory mediators, including the cytokines, among which interleukins 8 (IL-8) [16]. Determining which cytokines are implicated in the pathogenesis of EA may help in the diagnosis and treatment of this disease.

The aim of the study performed in EA-affected horses in asymptomatic, exacerbation and remission phase, is (1) to verify the diagnostic value of histology and the information deriving from it, using the histological scoring system for endoscopic biopsies, in a possible correlation with the clinical features of the different phases; (2) to analyze the immunohistochemical response for NKA and IL-8 from biopsy samples of lung tissue in subjects undergoing severe EA, in the different phases of experimentally induced disease; and (3) to evaluate the influence of sampling position along the respiratory tract on results.

2. Materials and Methods

2.1. Horses

Eleven horses were enrolled in the study, six EA-affected horses in asymptomatic phase, and five young just slaughtered horses as control, healthy at the pre-mortem visit and without evidence of pulmonary disease on post-mortem examination.

The EA-affected group consisted of five Italian Saddle horses and one Appaloosa, four mares and two geldings, 15.7 ± 1.9-year-old, with body weight ranging from 400 to 520 kg. The horses lived outdoor and had a history of signs of acute recurrent onset of asthma when housed in a stable with shavings for bedding and hay for feed. None of the horses received medications for at least three months prior to the assessment. Throughout the experimental time, a daily clinical examination was performed.

The control group comprised lungs removed immediately after slaughter from three Italian saddle horses and two Standardbreds, four mares and one male, 1.4 ± 1.5-year-old, with body weight ranging from 420 to 540 kg.

2.2. Experimental Design

EA-affected horses in asymptomatic phase (T0), were sedated with acepromazine maleate at 0.02 mg/kg bw iv (Prequillan, Fatro, Ozzano dell'Emilia BO, Italy) and detomidine hydrochloride at 0.01 mg/kg bw iv (Domosedan, Vetoquinol Italia, Bertinoro FC -Italy) and underwent a bronchoscopy (Pentax EG290P diameter 9.8 mm; length 120 cm).

For each endoscopic procedure performed in EA-affected horses, the instrument was introduced in the right and left principal bronchus. The airways were anesthetized by spraying of lidocaine, and four (two left and two right) epithelial biopsy specimens were obtained from four sites by use of endoscopic forceps. Biopsy sites were chosen between: 1) Right cranial bronchus and principal bronchus; 2) second right lateral segment and principal bronchus; 3) left cranial bronchus and principal bronchus; and 4) second left lateral segment and principal bronchus [24] (Figure 1).

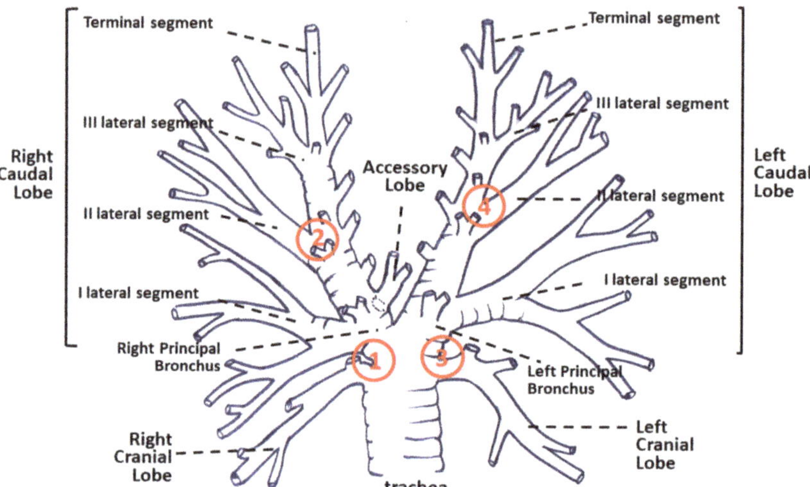

Figure 1. Position of bronchial biopsies: (1) Between right cranial bronchus and principal bronchus; (2) between second (II) right lateral segment and principal bronchus; (3) between left cranial bronchus and principal bronchus; and 4) between second (II) left lateral segment and principal bronchus.

Randomization of biopsy sequence was applied, and biopsy was performed by grabbing the lateral part of a bronchial branch with a 2.3 mm biopsy forceps, by avoiding the vessel from sliding flowing on the floor of the bronchus.

After the first bronchoscopy, the horses were brought from pasture, bedded in boxes with closed windows, on mold straw and fed hay of poor quality, to increase the dustiness of the environment and induce lower airway obstruction. Straw was not changed; it was rather raised twice a day to increase the dustiness of the environment. On day 2 (T1, early exacerbation phase) and 7 (T2, late exacerbation phase) of the challenge, bronchoscopy with bronchial biopsies was repeated with the same procedures described above. On day 8 the horses were put back to pasture after administration of dexamethasone (0.1 mg/kg) [25]. Four days later the biopsies were repeated (T3, remission phase) (Figure 2).

	T0	T1		T2		T3
Pasture		Box				Pasture
Day -3	0	2		7	8	12
Endoscopic biopsy	X	X		X		X

Figure 2. Timeline flow chart of trial.

For the control horses, the lung was removed as soon as the horses were slaughtered and placed on a flat surface. The presence of macroscopic lung changes was evaluated; the carina was opened to the bronchial branches, to acquire biopsy samples by using a biopsy punch in the same site chosen for EA-affected horses.

2.3. Bronchial Biopsy Histology

Immediately after biopsy collection, the samples were fixed in 10% neutral buffered formalin. Within 24–48 h, all samples were processed as routine for histology, embedded in paraffin, cut into 5 μm-thick sections and stained with hematoxylin and eosin (HE).

The observation of histological samples provided for an evaluation according to the standardized 14-point semiquantitative grading scoring system proposed by Bullone et al., 2016 for lung pulmonary biopsies based on histological features considered important in heaves [17]. The variables evaluated and scored were: (a) epithelial hyperplasia; (b) presence of epithelial inflammatory infiltrate; (c) goblet cells hyperplasia; (d) epithelial desquamation; (e) thickening of the basal membrane; (f) submucosal inflammation; (g) the presence of mucous glands within the lamina propria; (h) the presence of mucous gland among smooth muscle bundles; (i) airways smooth muscle fibrosis; and (j) presence of the end of the smooth muscle. Each variable was measured as 0 = absent and 1 = present, and the total (0–10) was calculated for each biopsy sample.

The biopsy specimens were scored by two veterinary pathologists at high and small magnification under a light microscope (NiKonEclipse 80, Nikon Corporation, Konan, Minato-ku, Tokyo, Japan). The microphotograph images were captured using a Nikon Digital Sight SD-MS camera (Nikon Corporation, Konan, Minato-ku, Tokyo, Japan) connected to the optical microscope.

2.4. Immunohistochemical Analysis for Neurokinin A and IL-8

Formalin fixed paraffin embedded samples were cut into 3–5 μm sections, mounted onto poly-L-lysine coated slides, dewaxed, rehydrated, and rinsed with tap water at room temperature. Immunolabelling was performed with a streptavidin–biotin–peroxidase technique (BIO SPA, Milan, Italy). The antibodies used and their dilutions were the following:NKA (1:400, polyclonal, Biorbyt, St. Louis, MO, USA) and IL-8 (1:100 dilution, monoclonal, Cloud Clone Corporation, Katy, TX, USA). The incubation was carried out with 3% hydrogen peroxide in methanol for 30 min (to block endogenous peroxidase activity) and microwave treatment (750 W) for antigen retrieval in citrate buffer solution at pH 6.0 (one burst of 5 min and five bursts of 2 min and 30 sec, replacing the evaporated buffer after each heating session). After that, the sections were incubated overnight at 4 °C in a humid chamber with the primary antibody diluted, according to the appropriate dilutions, in PBS (0.01 M, pH 7.4). The sections, subsequently washed in PBS, were incubated first with the secondary antibody (anti-rabbit IgG conjugated with biotin) for 30 min at room temperature, and then with the streptavidin-peroxidase complex for 25 min at room temperature. After a 12-min passage in the chromogenic DAB solution (diaminobenzidine 0.02%, and H_2O_2 0.001% in PBS), the sections were immediately rinsed in PBS, then in running water, stained with a contrast coloring (hematoxylin), dehydrated and assembled with DPX (Fluka, Riedel-de Häen, Germany).

NKA immunohistochemistry results were graded as follows: intensity of positivity (0, negative reaction; (1) weak intensity; (2) moderate intensity; (3) intense positivity; (4) very strong positivity); signal distribution (1, diffuse cellular, >50% of cells immunopositive; 2, outbreaks of positivity, <50% of the cells with a positive reaction), cell localization (1, cytoplasmic; 2: nuclear; 3: both cytoplasmic and nuclear).

IL-8 immunohistochemistry results were graded as follows: intensity of positivity (0, negative reaction; 1, weak intensity; 2, moderate intensity; 3, intense positivity; 4, very strong positivity); signal distribution (1, diffuse cellular, >50% of cells immunopositive; 2, outbreaks of positivity, <50% of the cells with a positive reaction).

Appropriate positive controls were used in order to evaluate the specificity of the reactions and ascertain the proper cross-reactivity in the horse tissue. Brain and normal horse lung were used as positive controls for NKA, horse lymph node as IL-8 control. As a negative control for the immunohistochemical procedure, 10% normal mouse serum was used in replicate sections instead of the primary antibody.

All the markers were evaluated through blinded observations by semiquantitative analysis of five representative high-power fields at the optical microscope (NiKonEclipse 80, Nikon Corporation, Japan). The microphotograph images were captured using a Nikon Digital Sight SD-MS camera (Nikon Corporation, Konan, Minato-ku, Tokyo, Japan) connected to the optical microscope.

2.5. Statistical Analysis

Statistical analysis was performed with a commercially available program (GraphPad Prism 5, GraphPad Software, San Diego, CA, USA). Assessment of data for normality was calculated by applying the D'Agostino-Pearson test. Data were expressed as median (minimum-maximum).

Differences in histological score, NKA and IL-8 immunohistochemistry between four bronchial sampling sites were analyzed by a Kruskal–Wallis one-way analysis of variance, both in EA-affected horses (at each experimental time), and in control horses.

For each parameter examined in each experimental time, the median value of variables of four biopsy samples was used for the subsequent statistical analysis.

A Mann–Whitney test (two tail p value) was applied to perform a comparison between control horses and EA-affected horses, at each experimental time, for the following variables: (1) Histological score; (2) NKA immunohistochemistry score; and (3) IL-8 immunohistochemistry score. Similarly, a Friedman test with Dunn's multiple comparison test, as post hoc test, was applied to perform a comparison of EA-affected horses at different experimental times, for the following variables: (1) Histological score; (2) NKA immunohistochemistry score; and (3) IL-8 immunohistochemistry score.

The significance was set for $p < 0.05$.

3. Results

All the horses, initially asymptomatic, showed respiratory signs related to equine asthma exacerbation, after the transition from the pasture to the stable. Indeed, after 2 days from the beginning of the test (T1), clinical examination showed cough and dyspnea in all subjects. However, dyspnea showed a quick remission after environmental change and drug treatment at T3.

Histological score, and NKA and IL-8 immunohistochemistry scores of bronchial biopsies revealed no significant differences between the different biopsy sites nor in EA-affected horses in every experimental time, and in slaughtered horses (Table 1).

With the exception of the first sampling performed, endoscopic biopsies in EA-affected horses were found to be evaluable in at least one of the four samples each horse in each phase, and therefore samples of all phases in all cases were included in the statistical evaluation, both for histology and for immunohistochemistry.

The histopathology evaluation of the bronchial samples collected in the slaughtered horses ranged from a score of 2–3, allowing to consider those subjects as not affected by any respiratory problem, and to include them all as control-cases (see Supplementary Material Table S1). The histological score of EA-affected horses is detailed in Supplementary Material, Table S1. Relevant pictures of histological features are reported in Figure 3.

The comparison of histological scores of control vs EA-affected horses at four experimental points (T0; T1; T2; T3) demonstrated a significant difference at T0 ($p = 0.001$), in the asymptomatic phase; at T1 ($p = 0.007$), in the early exacerbation phase; at T2 ($p = 0.007$), in the late exacerbation phase; and at T3 ($p = 0.007$), in the remission phase (Table 2).

Table 1. Comparison of histological, NKA immunoreactivity score and IL-8 immunoreactivity score acquired from bronchial biopsies, obtained in four standardized position, from control horses and EA-affected horses at T0, T1, T2, T3. Median value (min-max) and *p* value—Kruskal–Wallis test.

	Control	EA-Affected	EA-Affected	EA-Affected	EA-Affected
		T0	T1	T2	T3
Histological score (0–10)	3	5.5	5	6.5	7.5
median (min-max)	(2–4)	(4–6)	(4–8)	(5–10)	(5–9)
p	$p = 0.92$	$p = 0.48$	$p = 0.16$	$p = 0.63$	$p = 0.81$
NKA immunoreactivity score					
Intensity of positivity (0–4)	2	3	3	3	4
median (min-max)	(2–3)	(1–4)	(2–4)	(2–4)	(2–4)
p	$p = 0.5$	$p = 0.88$	$p = 0.49$	$p = 0.83$	$p = 0.06$
Signal distribution (1–2)	1	1	2	1	1
median (min-max)	(1–1)	(1–2)	(1–2)	(1–2)	(1–2)
p		$p = 0.27$	$p = 0.99$	$p = 0.54$	$p = 0.91$
Cell localization (1–3)	1	1.5	1	1.5	2
median (min-max)	(1–1)	(1–3)	(1–3)	(1–3)	(1–3)
p		$p = 0.58$	$p = 0.96$	$p = 0.41$	$p = 0.85$
IL-8 immunoreactivity score					
Intensity of positivity (0–4)	2	3	3	3	3
median (min-max)	(1–3)	(3–4)	(2–4)	(2-4)	(2-4)
p	$p = 0.07$	$p = 0.65$	$p = 0.87$	$p = 0.35$	$p = 0.34$
Signal distribution (1–2)	2	1	2	1	1
median (min-max)	(1–2)	(1–2)	(1–2)	(1–2)	(1–2)
p	$p = 0.5$	$p = 0.56$	$p = 0.63$	$p = 0.48$	$p = 0.41$

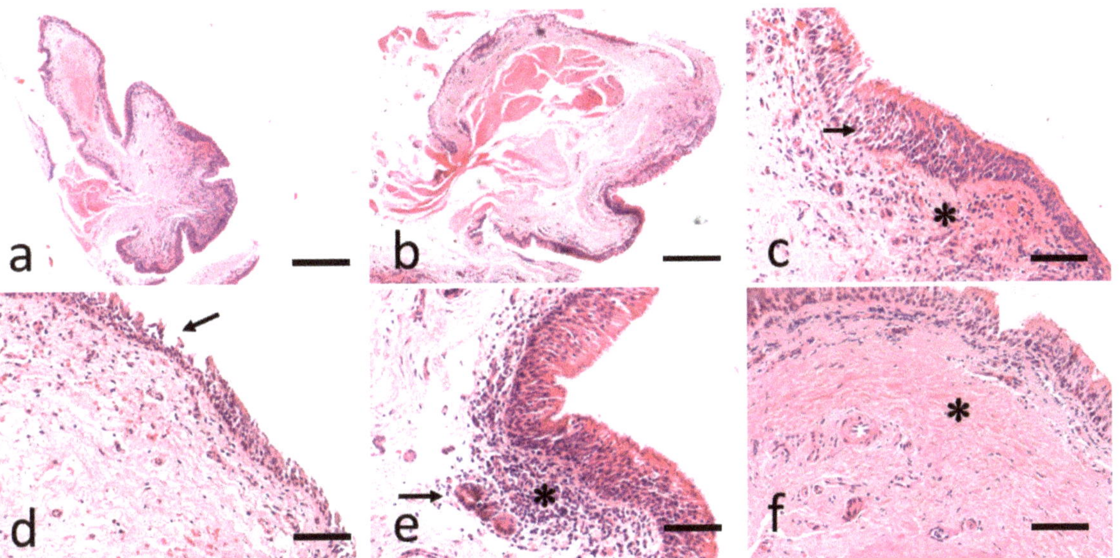

Figure 3. Examples of histological samples of bronchial biopsy in the EA horses. (**a**,**b**). Low-magnification images of biopsies, obtained with tissue twisting technique with biopsy forceps before tearing. All tissue layers are observed (surface epithelium, extracellular matrix and airway smooth muscle). HE, bar 800 μm. (**c**) Marked mucosal hyperplasia (arrow), and the presence of a moderate inflammatory submucosal infiltrate of lymphocyte (asterisk). HE, bar 200 μm. (**d**) Epithelial desquamation involving over 70% of the mucous layer (arrow). HE, bar 200 μm. (**e**) Severe diffuse lymphocytic infiltrate (asterisk) and presence of mucosal gland (arrow). HE, bar 200 μm. (**f**) Severe submucosal fibrosis (asterisk). HE, bar 200 μm.

Table 2. Comparison of median value (min-max) of histological, NKA and IL-8 immunoreactivity scores acquired from bronchial biopsies, between control horses and EA-affected horses at T0, T1, T2, T3. Mann–Whitney test. a,b: different letters indicate significant differences ($p < 0.05$); A, B: different letters indicate significant differences ($p < 0.01$).

	Control	EA-Affected	EA-Affected	EA-Affected	EA-Affected
		T0	T1	T2	T3
Histological score (0–10)	3	5.5	5	6.5	7.5
median (min-max)	(2–4) A	(4–6) B	(4–8) B	(5–10) B	(5–9) B
NKA immunoreactivity score					
Intensity of positivity (0–4)	2	3	3	3	4
median (min-max)	(2–3) a	(1–4)	(2–4)	(2–4) b	(2–4) b
Signal distribution (1–2)	1	1	2	1	1
median (min-max)	(1–1)	(1–2)	(1–2)	(1–2)	(1–2)
Cell localization (1–3)	1	1.5	1	1.5	2
median (min-max)	(1–1)	(1–3)	(1–3)	(1–3)	(1–3)
IL-8 immunoreactivity score					
Intensity of positivity (0–4)	2	3	3	3	3
median (min-max)	(1–3)	(3–4)	(2–4)	(2–4)	(2–4)
Signal distribution (1–2)	2	1	2	1	1
median (min-max)	(1–2)	(1–2)	(1–2)	(1–2)	(1–2)

Median histological scores of bronchial biopsies of EA-affected horses during the different experimental phases (T0, T1, T2, T3), evidenced a significant increase moving from the asymptomatic phase (T0) to the remission phase (T3) with significant difference between T0 and T3 ($p = 0.01$) (Table 3).

Table 3. Comparison of median value (min-max) of histological, NKA and IL-8 immunoreactivity scores acquired from bronchial biopsies, between EA-affected horses at T0, T1, T2, T3. Friedman test with Dunn's multiple comparison test as post hoc test. a,b: different letters indicate significant differences ($p < 0.05$).

	EA-Affected	EA-Affected	EA-Affected	EA-Affected
	T0	T1	T2	T3
Histological score (0–10)	5.5	5	6.5	7.5
median (min-max)	(4–6) a	(4–8)	(5–10)	(5–9) b
NKA immunoreactivity score				
Intensity of positivity (0–4)	3	3	3	4
median (min-max)	(1–4)	(2–4)	(2–4)	(2–4)
Signal distribution (1–2)	1	2	1	1
median (min-max)	(1–2)	(1–2)	(1–2)	(1–2)
Cell localization (1–3)	1.5	1	1.5	2
median (min-max)	(1–3)	(1–3)	(1–3)	(1–3)
IL-8 immunoreactivity score				
Intensity of positivity (0–4)	3	3	3	3
median (min-max)	(3–4)	(2–4)	(2–4)	(2–4)
Signal distribution (1–2)	1	2	1	1
median (min-max)	(1–2)	(1–2)	(1–2)	(1–2)

The results for each immunohistochemical marker are reported in Tables 4 and 5. Relevant pictures of the IHC on control and EA-affected horses for NKA and IL-8 are reported in Figures 4 and 5, respectively.

Table 4. Results of immunohistochemical evaluation to NKA of all endoscopic samples in the six EA-affected horses in all phases of experimental trial (T0, T1, T2, T3).

NKA	EA-Case 1				EA-Case 2				EA-Case 3				EA-Case 4				EA-Case 5				EA-Case 6			
	B Right		B Left		B Right		B Left		B Right		B Left		B Right		B Left		B Right		B Left		B Right		B Left	
	P	D	P	D	P	D	P	D	P	D	P	D	P	D	P	D	P	D	P	D	P	D	P	D
T0	n.e.	n.e.	n.e.	n.e.	2, N, diffuse	2, C, diffuse	3, N, diffuse	2, C, diffuse	n.e.	n.e.	n.e.	n.e.	2, C, diffuse	3, CN, diffuse	n.e.	2, N, diffuse	3, C, focal	n.e.	n.e.	3, CN, diffuse	1, C, focal	n.e.	1, C, diffuse	n.e.
T1	n.e.	2, C, diffuse	n.e.	2, C, diffuse	1, C, diffuse	3, N, diffuse	2, C, diffuse	3, CN, diffuse	2, CN, diffuse	3, CN, diffuse	2, CN, diffuse	3, CN, diffuse	1, C, focal	1, C, focal	1, C, focal	1, C, focal	n.e.	3, C, focal	3, C, focal	3, C, focal	1, CN, focal	2, CN, focal	3, CN, focal	2, C, focal
T2	n.e.	2, C, diffuse	n.e.	3, CN, diffuse	2, C, diffuse	2, C, diffuse	2, C, diffuse	1, C, diffuse	2, C, diffuse	3, CN, diffuse	2, C, diffuse	3, CN, diffuse	1, C, focal	2, CN, diffuse	2, C, diffuse	3, CN, diffuse	2, CN, focal	3, CN, diffuse	2, C, focal	3, N, focal	2, CN, focal	1, CN, focal	1, C, focal	1, C, focal
T3	3, N, diffuse	3, N, diffuse	3, N, diffuse	2, N, diffuse	n.e.	3, CN, diffuse	3, CN, diffuse	1, C, diffuse	3, CN, diffuse	3, CN, diffuse	3, C, diffuse	3, C, diffuse	3, CN, focal	n.e.	3, CN, diffuse	3, CN, diffuse	3, CN, diffuse	3, CN, focal	3, N, focal	2, CN, focal	1, C, focal	n.e.	n.e.	1, C, focal

B: bronchus, P: proximal, D: distal; C: cytoplasmic immunopositivity; N; nuclear immunopositivity; n.e.; not evaluable.

Table 5. Results of immunohistochemical evaluation to IL-8 of all endoscopic samples in the six EA-affected horses in all phases of experimental trial (T0, T1, T2, T3).

IL-8	EA-Case 1				EA-Case 2				EA-Case 3				EA- Case 4				EA- Case 5				EA- Case 6			
	B Right		B Left		B Right		B Left		B Right		B Left		B Right		B Left		B Right		B Left		B Right		B Left	
	P	D	P	D	P	D	P	D	P	D	P	D	P	D	P	D	P	D	P	D	P	D	P	D
T0	3, focal	n.e.	3, focal	n.e.	3, diffuse	4, diffuse	4, diffuse	2, diffuse	n.e.	n.e.	n.e.	n.e.	3, focal	4, diffuse	n.e.	3, diffuse	4, diffuse	3, focal	n.e.	3, diffuse	n.e.	n.e.	n.e.	n.e.
T1	n.e.	2, focal	n.e.	3, focal	3, diffuse	3, focal	3, focal	4, diffuse	n.e.	3, diffuse	4, diffuse	3, focal	2, focal	3, focal	2, focal	3, focal	4, diffuse	4, diffuse	4, diffuse	4, diffuse	4, diffuse	4, focal	3, focal	4, diffuse
T2	n.e.	4, diffuse	3, diffuse	3, focal	3, diffuse	3, diffuse	3, focal	4, diffuse	2, focal	2, diffuse	2, focal	2, diffuse	2, focal	3, diffuse	3, diffuse	3, focal	3, diffuse	3, diffuse	4, focal	2, diff	2, focal	4, focal	3, diffuse	2, focal
T3	4, diffuse	4, diffuse	3, diffuse	3, focal	n.e.	n.e.	4, diffuse	4, diffuse	3, diffuse	4, diffuse	3, diffuse	4, diffuse	4, diffuse	n.e.	3, diffuse	3, diffuse	4, diffuse	4, focal	4, focal	n.e.	n.e.	3, focal	2, focal	2, focal

B: bronchus, P: proximal, D: distal; n.e.; not evaluable.

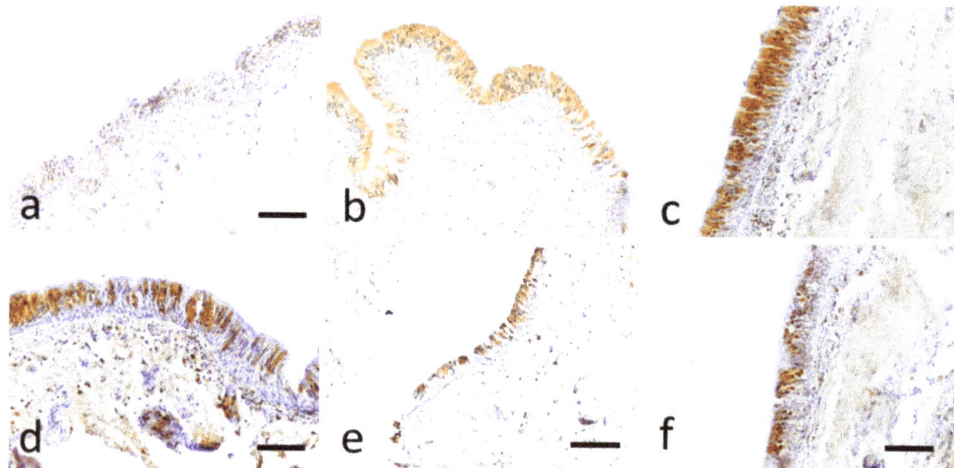

Figure 4. NKA immunohistochemical stain (brown color), bar 200 µm. (**a**) Bronchial mucosa of a control horse. Weak and diffuse cytoplasmic positivity of the respiratory epithelium. (**b**) EA-case 3, distal right bronchus, T2. Moderate and diffuse cytoplasmic positivity to the epithelial cells of the respiratory mucosa, associated with an intense nuclear positivity. (**c**) EA-case 1 horse, proximal right bronchus, T3. Intense, diffuse, cytoplasmic and nuclear positivity, mainly localized in the more superficial layers of the hyperplastic respiratory epithelium. (**d**) EA-case 3 horse, distal right bronchus, T3. Intense, diffuse cytoplasmic and nuclear localized mainly in the superficial layers of the epithelium. (**e**) EA-affected 6, proximal left bronchus, T1. Positivity confined to the luminal layer of the epithelium. (**f**) EA-case 2, distal left bronchus, T3. Predominantly nuclear positivity.

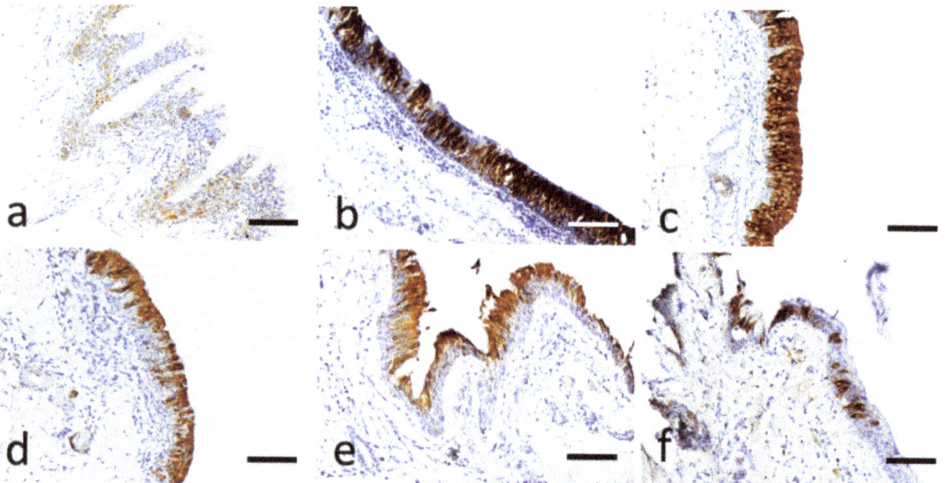

Figure 5. IL-8 immunohistochemical stain (brown color), bar 200 µm. (**a**) Bronchial mucosa of a control horse. Focal and weak cytoplasmic positivity to the respiratory epithelium. (**b**) EA-case 2, distal right bronchus, T3. Strong and diffuse positivity to the epithelial cells of the mucosa. (**c**) EA-case 2, proximal right bronchus, T3. Note the intense immunopositivity in all layers of the hyperplastic mucosa. (**d**) EA-case 2, distal right bronchus, T2. Note the distribution of positivity in the most superficial layers of the mucosa. (**e**) EA-affected horse 6, proximal left bronchus, T3. The positivity appears to be confined to the most superficial layer of the epithelium also in this case. (**f**) EA-case 5, proximal left bronchus, T1. Focal positivity confined to isolated respiratory epithelial cells.

In control horses, NKA immunolabelling was weak, widespread cytoplasmic in the epithelial bronchial cells. In EA-affected horses, NKA immunolabelling appeared moderate to strong staining at the cytoplasm and occasionally in the nucleus (was found in a minority of epithelial cells also in the nucleus) (Figure 4c,d). In many cases, the immunopositivity was confined to the apical portion of epithelial cells of the airway mucosa (Figure 4e).

The expression of IL-8 was cytoplasmatic. IL-8 positivity was seen in epithelial cells of control horses with a discontinuous, focal, weak staining. IL-8 immunolabelling in equine asthma instead appeared multifocal to diffuse, cytoplasmic, and moderate to intense (Figure 5b,c), in some cases confined to isolated respiratory epithelial cells (Figure 5e,f).

NKA immunopositivity scores of bronchial biopsies showed a significant increase of intensity of positivity, moving from control horses to EA-affected horses at T2 ($p = 0.04$), and at T3 ($p = 0.04$), while no significant differences were showed for signal distribution and cell localization between groups (Table 2). No differences for NKA immunohistochemistry results (intensity of positivity, signal distribution, cell localization), were recorded in EA-affected horses, among the experimental times (Table 3).

At last, IL-8 immunopositivity score, subdivided on intensity of positivity and signal distribution, recorded no significant differences between control horses and EA-affected horses (Table 2), nor in EA-affected horses in each experimental time (Table 3).

4. Discussion

The diagnosis of severe equine asthma (EA) is mainly based on history and clinical signs [2,3,6,25]. Collateral diagnostic investigations with additional tests, such as an airway endoscopy or lung function evaluation, can confirm and further characterize the diagnostic suspect of severe equine asthma. Among these additional tests, the cytological examination of bronchoalveolar lavage fluid (BALF) is still considered as a 'gold standard' in the diagnosis of severe lower airway inflammation in the horse [2,6,16,18].

However, if on the one hand the BALF cytology is able to disclose neutrophilic inflammation during the exacerbation of the disease [14], on the other hand, the inflammatory pattern completely normalizes during periods of remission, induced by antigen avoidance strategies [17,26]. Clinical signs are also largely reversible after treatment with inhaled or systemic corticosteroids, in absence of harmful environment [2,17]. Conversely, remodeling of the peripheral airways in the severe EA persists during the remission phases and it is related to residual airflow obstruction [26].

Airway remodeling includes increased airway smooth muscle mass, goblet cell hyperplasia/metaplasia, collagen and elastic fiber deposition within the lamina propria, peribronchiolar fibrosis, airway obstruction by mucus and inflammatory cells, airway adventitial inflammation [17,27]. Therefore, it is not yet clear which role the above-mentioned remodeling process plays and how much it may affect airways inflammation and especially in maintaining disease during asymptomatic periods. Moreover, due to the absence of a significant association between lower airway inflammation detected with BALF and the degree of pulmonary dysfunction in equine asthma, the study of the structural small and large airway alterations has recently gained interest [2]. Identifying specific pathologic variables by endoscopic biopsy specimens could possibly facilitate the reaching of a diagnosis during the remission and monitoring response to treatment [17].

The histopathological characteristics of the bronchial mucosa and their evaluation in the diagnosis of severe EA are currently poorly characterized and very few information concerning the histological evaluation of endobronchial biopsies of the horse is present in the literature [17,18,28]. The main reasons for this lack consist of the fact that the transthoracic biopsy samples are a risky and expensive procedure, and therefore not carried out in clinical practice. The only studies existing in the literature for the histological evaluation of peripheral airway remodeling concern histological studies on lungs of horses during autopsy or slaughter [27,29]. On the contrary, the only method that can be used in life is an endobronchial biopsy sampling from the central airways, which is quite easy and repeatable. Nevertheless, the biopsies obtained from the large respiratory tract do not specifically collect material from the areas affected by the pathology; the information is still very scarce in the course of EA [2]. If the effects were similar, you could then be able to make a diagnosis during the clinical remission phase of the severe EA. Furthermore, the studies carried out on BALF and on endobronchial biopsy specimens have focused on biomolecular type investigations gene expression [5,30–40]. Contrariwise, only very few papers deal with biopsies samples on paraffin sections for histological and immunohistochemical evaluations [5,17,18,20,37].

In our procedure, the difficulties in bioptic procedures encountered in the first sampling (EA-affected 3), suggested by the classic methodologies, which made the sampled material too superficial and unsuitable for evaluation, were largely solved using a manual technique that provided for access to the wall of the bronchus, by withdrawing the forceps towards the endoscope and at the same time by exerting a twist in such a way as to acquire a larger sample.

Bullone et al., 2016 first developed a 14-point histological scoring system that assesses histological variables of inflammation and remodeling of the large airways using endoscopic biopsy specimens [17]. This evaluation system, however, although considered a reliable tool for the assessment of airway obstruction caused by inflammation and remodeling in horses with severe forms of EA, still requires a standardized evaluation and is currently rarely used [3,18]. Their application of this scoring system in 20 horses (controls, exacerbation and remission of heaves) disclose a correlation between histology and the degree of airway obstruction clinically measured, but does not seem to discriminate horses with heaves in remission from control [17].

Using a classical histologic assessment for evaluation of endoscopic biopsies, Niedzwieds et al., 2018 also found no significant difference for any of the histological variables between control and study group with asthma, and concludes hoping for the use of a standardized scoring system, such that developed by Bullone et al., 2016, to differentiate horses with an asthma exacerbation from healthy horses [17,18].

In our study, we chose to use the same scoring system to Bullone et al., 2016 for histologic evaluation of biopsy samples [17]. Conversely to previously results, we found significant differences in histological scoring between subjects in remission and healthy control horses. These results demonstrate how, similarly to what happens for the remodeling of the peripheral airways, the progression and maintenance of the disease are preserved histologically even in the remission phase, and an altered immunoregulatory function of the airway epithelium in horses with severe equine asthma syndrome.

The most common mediators of airway inflammation involved in horse respiratory diseases include histamine, bradykinin, prostaglandins, leukotrienes, platelet factor and endothelin-1 [17,40,41]. Furthermore, another mediator with a physiological and pathological role in respiratory function is a neuropeptide called neurokinin A (NKA), a member of the tachykinin family [42–44]. Neurokinin A is involved in nonadrenergic/noncolinergic neurotransmitter mediation of excitatory type in the respiratory tract of rat and guinea pig in normal conditions [45,46]. The NKA effect manifests through the activation of NK-2 receptors; they contract the smooth muscle of the airways leading to narrowing of the lumen. An abnormal expression of NK-2 receptors has been associated in human beings with forms of asthma [47,48].

On the one hand, the role of NKA in bronchoconstriction and neurogenic inflammation in patients with asthma generated scientific attention and specific investigations on selective NKA receptor antagonists for therapeutic uses [49]. On the other hand, Neurokinin B has been shown to be a much less potent contractile agent in vitro than NKA in both control and EA horses, probably due to fewer receptors or less affinity [20]. However, little information exists regarding the involvement of NKA receptors in the horse, moreover limited to the intestine [50] and to the lung of horses affected by EA [20]. Recently, the receptors of the capsaicin sensitive-sensory nerves stimulated by bacterial of lipopolysaccharides, capable of inducing neurogenic inflammation [23] as occur in human airways [48] have been investigated in equine bronchial tissue.

The research conducted by Venugopal et al., 2009 on healthy and EA-affected horses through a comparative method demonstrates how significantly the expression of NK-2 has increased in the bronchial epithelium and in the bronchial smooth muscle and lung vessels of EA-affected horses [20]. The results of this study demonstrate how much NK-2 receptors are upregulated in EA, and suggest that NK-2 receptor antagonists may have some therapeutic effect in controlling the progression of airway hyperreactivity. Based on the theory that both NKA and NKB can cause concentration-dependent contractions in horse bronchial rings, and that this response is more evident in horses affected by severe equine asthma, the Authors hypothesized that the increase in the expression of NK-2 receptors may be a major cause of airway hyperreactivity in EA-affected horses.

The ex vivo study by Calzetta et al., 2018 confirms and extends the previous results by Venugopal et al., 2009 and suggests that the environmental exposure to bacterial of lipopolysaccharides may represent a crucial factor in modulating the bronchial responsiveness of severe equine asthma [20,23]. A chronic exposure to high concentration of inhaled endotoxin in fact has a deleterious effect in equine distal airways: it triggers dysfunctional airway smooth muscle (ASM) contractility due to the stimulation of capsaicin-sensitive sensory nerves, increased the release of NKA and the activation of NK2 receptors. Antagonizing NK2 receptors may have a beneficial impact on normalizing the contractility and on controlling clinical signs in severe equine asthma.

Our results reveal a significant variability in the immunohistochemical expression of NKA between control horses and horses with severe equine asthma in remission. However, contrarily to what was expected, NKA immunoreactivity appears to increase significantly after corticosteroid treatment. This demonstrates to what extent the clinical improvement associated with therapy does not correspond to a decrease in this marker of inflammation, but rather show a contrary trend. Therefore, this result could suggest that, unlike cortisone drugs, there are potential bases for an efficacy of selective and non-selective NK2 receptor treatments.

T lymphocytes play a fundamental role in modulating the immune response during the pathogenesis of severe equine respiratory syndrome. Data of the literature suggests that lung T helper may be implicated through the secretion of Th-1 or Th-2 type cytokines [19,37,38,51–54]. Horses affected by severe equine asthma syndrome produce both of these cytokines, depending on the stage of the disease and the time when sampling is performed [55]. In addition, the expression of cytokines in the airway lymphocytes also appears to be influenced by the length of time in which the horses have manifested the disease clinically [55].

The bronchiolar intraluminal accumulation of neutrophils usually accompanies the lymphocyte peribronchiolar infiltrate in severe EA anatomopathological findings [56], and takes place a few hours after environmental changes. The late phase of type I hypersensitivity plays a central role in a Th mixed response, mainly Th2 at the beginning and Th1 in the late phase of the process [35]. Both ways contribute to triggering inflammation and neutrophil recruitment in the airways by means of several cytokines, such as IL-8 [37,52,57]. In addition, a type III hypersensitivity reaction is able to partially explain the neutrophilic inflammation in the airways on EA-affected horses. However, the factors that initiate this process are still not fully explained [55].

Studies on horses with severe equine respiratory syndrome have revealed an increase in the gene expression of the pro-inflammatory cytokine IL-8 in bronchoalveolar cells [19,38,40,52,58,59], as well as an increase in IL-8 protein concentration in BALF [19,37,38,40,60–65].

It is well known that the main producers of IL-8 are the airway epithelial cells [27]. Ainsworth et al., 2006 demonstrated a 3- to 10-fold increase of IL-8 m-RNA in epithelial airway cells in severe equine asthma, and afterwards Berndt et al., 2007 confirmed how IL-8 mRNA expression in bronchial epithelial cells increases after exposure to stable dust, both in EA-affected subjects and in control animals [27,61]. These results were also confirmed by another study [38], in which an increase in the level of IL-8 mRNA is observed both in BAL cells and in endobronchial biopsies during the crisis phases of the disease.

On the contrary, in the study by Pietra et al., 2011, although a significant overexpression of IL-8 and TNFα mRNA on BALF was observed in the group of treated horses, with a peak around the ninth day, a comparison between exposed and unexposed horses does not reveal significant differences [64]. According to the authors, this evidence could depend on the loss of uniformity of the sample, since the evolution of the pathology represents a continuum from healthy to pathological conditions.

Padoan et al., 2013, in a comparative research between clinical and endoscopic findings, cytological and cultural tests of BALF, histology of bronchial tissue and analysis of gene expression of inflammatory mediators in BALF and biopsies, found that six out of ten immunologically related genes (including IL-8) show a significant difference in expression between horses affected by severe equine asthma syndrome and the control group [40]. Specifically, however, IL-8 mRNA levels measured in biopsies are a hundred times lower than in BALF, and although an increase in its gene expression has been found in bronchial biopsies, this difference is not statistically relevant. Furthermore, the biopsies included in the study showed no significant differences in gene expression levels between EA-affected horses and controls. The lack of statistical significance between mRNA levels of inflammatory mediators in the respiratory epithelium could be caused by the small size of the samples or by the site where the biopsy is performed.

By using immunohistochemistry, Ainsworth et al., 2006 demonstrated IL-8 localization in the cytoplasm of epithelial cells in airway biopsies as well as in more recent times Tessier et al., 2017 observed a marked increase in IL-8 gene product determined by immunohistochemistry in bronchial cells in asthmatic, but not in non-asthmatic animals. In our study, we found significant differences between healthy and EA-affected in epithelial immunohistochemical expression, but none in IL-8 immunohistochemistry expression in the different stages of EA [5,37]. Unlike former suggestions in literature data on biomolecular BALF evaluation, however, this result appears to be aligned with the histological evaluations, which do not reveal any relevant difference in the inflammatory evaluation parameters in all phases.

Beyond the statistical significance of the results, the immunohistochemical observations reveal however interesting suggestions: the reactivity to NKA shows a trend to increase in intensity and to show same dysregulation characteristics (with the appearance of abnormal nuclear locations and variations in the mucosal distribution) in horses with severe equine syndrome compared to those of control. For IL-8, in horses with EA there is an intense positivity—often focal and localized to in single cells—which instead is not present in all the control subjects, where positivity is always widespread and of far weaker intensity. As far as these observations have diagnostic and/or immunopathogenic relevance for this disease in its various stages, it is not currently possible to determine it with certainty. Moreover, studies in this regard require a number of samples significantly higher and accompanied by biomolecular analyzes.

5. Conclusions

The results of the present study demonstrate homogeneity in the bronchial endoscopic samples regardless of the sampling site, demonstrating a lack of influence of bronchial sampling position on histological and immunohistochemical evaluation. The application of the histological score of Bullone et al., 2016 to endoscopic biopsies [17], however, fails to first differentiate between the exacerbation and the remission phase, which, contrarily to what we found, should have a return to histological conditions significantly similar to those of normality.

By immunohistochemical analysis, the comparison between clinically healthy horses and EA-affected ones suggested new insights on the cytokine expression in equine health and disease status. On the one hand, investigations on endoscopic biopsy samples allowed detecting relevant differences in immunoreactivity between control horses and horses with severe equine asthma syndrome, demonstrating a dysregulation of their expression in the latter. On the other hand, the clinical improvement at different times of the disease seems to have no immunohistochemical bases for IL-8 and NKA expression. The explanations behind these results are still being studied and require more data and subjects examined.

Supplementary Materials: The following are available online at https://www.mdpi.com/article/10.3390/ani11051376/s1, Table S1: Excel file reporting all the histological data collected for this study.

Author Contributions: Conception and design of the work: M.M., M.P., A.P., N.R., and A.S.; acquisition, analysis, and interpretation of data: M.M., M.P., R.R., and G.M.; drafted the work: M.M. and M.P.; substantively revised the manuscript: M.M., M.P., A.P., G.M., N.R., and R.R. All authors have read and agreed to the published version of the manuscript.

Funding: This research has no external funding.

Institutional Review Board Statement: The study was approved by the Ethics and Scientific Committee of the Alma Mater Studiorum University of Bologna in the plenary session on 3 March 2009 (protocol n° 04/55/09) and submitted to the Italian Ministry of Health. All experimental procedures were carried out in accordance with European legislation regarding the protection of animals used for experimental and other scientific purposes (Council Directive 86/609/EEC).

Informed Consent Statement: Informed consent was obtained from all subjects involved in the study.

Data Availability Statement: All data generated or analyzed during this study are included in this published article. The raw datasets used and analyzed during the current study are available from the corresponding author on reasonable request.

Acknowledgments: Authors thank Laura Joanna Luczak, English translator and interpreter, for the English revision of the manuscript.

Conflicts of Interest: The authors declare no conflict of interest.

References

1. Martin, B.B., Jr.; Reef, V.B.; Parente, E.J.; Sage, A.D. Causes of poor performance of horses during training, racing, or showing: 348 cases (1992–1996). *J. Am. Vet. Med. Assoc.* **2000**, *216*, 554–558. [CrossRef] [PubMed]
2. Bullone, M.; Lavoie, J.P. Science-in-brief: Equine asthma diagnosis: Beyond bronchoalveolar lavage cytology. *Equine Vet. J.* **2017**, *49*, 263–265. [CrossRef] [PubMed]
3. Couetil, L.; Cardwell, J.M.; Leguillette, R.; Mazan, M.; Richard, E.; Bienzle, D.; Bullone, M.; Gerber, V.; Ivester, K.; Lavoie, J.P.; et al. Equine Asthma: Current Understanding and Future Directions. *Front. Vet. Sci.* **2020**, *7*, 450. [CrossRef]
4. Couetil, L.L.; Cardwell, J.M.; Gerber, V.; Lavoie, J.P.; Leguillette, R.; Richard, E.A. Inflammatory airway disease of horses–revised Consensus statement. *J. Vet. Intern. Med.* **2016**, *30*, 503–515. [CrossRef] [PubMed]
5. Tessier, L.; Côté, O.; Clark, M.E.; Viel, L.; Diaz-Méndez, A.; Anders, S.; Bienzle, D. Impaired response of the bronchial epithelium to inflammation characterizes severe equine asthma. *BMC Genom.* **2017**, *18*, 708. [CrossRef]
6. Bond, S.; Léguillette, R.; Richard, E.A.; Couetil, L.; Lavoie, J.P.; Martin, J.G.; Pirie, R.S. Equine asthma: Integrative biologic relevance of a recently proposed nomenclature. *J. Vet. Int. Med.* **2018**, *32*, 2088–2098. [CrossRef]
7. Seahorn, T.L.; Beadle, R.E. Summer pasture-associated obstructive pulmonary disease in horses: 21 cases (1983–1991). *J. Am. Vet. Med. Assoc.* **1993**, *202*, 779–782.
8. McGorum, B.C.; Ellison, J.; Cullen, R.T. Total and respirable airborne dust endotoxin concentrations in three management systems. *Equine Vet. J.* **1998**, *30*, 430–434. [CrossRef]
9. Vandenput, S.; Votion, D.; Duvivier, D.H.; van Erck, E.; Anciaux, N.; Art, T.; Lekeux, P. Effect of a set stabled environmental control on pulmonary function and airway reactivity of COPD affected horses. *Vet. J.* **1998**, *155*, 189–195. [CrossRef]
10. Tremblay, G.M.; Ferland, C.; Lapointe, J.M.; Vrins, A.; Lavoie, J.P.; Cormier, Y. Effect of stabling on bronchoalveolar cells obtained from normal al COPD horses. *Equine Vet. J.* **1993**, *25*, 194–197. [CrossRef]
11. Pirie, R.S. Recurrent airway obstruction: A review. *Equine Vet. J.* **2014**, *46*, 276–288. [CrossRef]
12. Fogarty, U.; Buckley, T. Bronchoalveolar lavage findings in horses with exercise intolerance. *Equine Vet. J.* **1991**, *23*, 434–437. [CrossRef] [PubMed]
13. Holcombe, S.J.; Jackson, C.; Gerber, V.; Jefcoat, A.; Berney, C.; Eberhardt, S.; Robinson, N.E. Stabling is associated with airway inflammation in young Arabian horses. *Equine Vet. J.* **2001**, *33*, 244–249. [CrossRef] [PubMed]
14. Jean, D.; Vrins, A.; Beauchamp, G.; Lavoie, J.P. Evaluation of variations in bronchoalveolar lavage fluid in horses with recurrent airway obstruction. *Am. J. Vet. Res.* **2011**, *72*, 838–842. [CrossRef]
15. Hoffman, A.M. Bronchoalveolar lavage: Sampling technique and guidelines for cytologic preparation and interpretation. *Vet. Clin. N. Am. Equine Pract.* **2008**, *24*, 423–435. [CrossRef] [PubMed]

16. Couetil, L.L.; Thompson, C.A. Airway Diagnostics: Bronchoalveolar Lavage, Tracheal Wash, and Pleural Fluid. *Vet. Clin. N. Am. Equine Pract.* **2020**, *36*, 87–103. [CrossRef] [PubMed]
17. Bullone, M.; Helie, P.; Joubert, P.; Lavoie, J.P. Development of a Semiquantitative Histological Score for the Diagnosis of Heaves Using Endobronchial Biopsy Specimens in Horses. *J. Vet. Intern. Med.* **2016**, *30*, 1739–1746. [CrossRef]
18. Niedzwiedz, A.; Mordak, R.; Jaworski, Z.; Nicponc, J. Utility of the Histological Examination of the Bronchial Mucosa in the Diagnosis of Severe Equine Asthma Syndrome in Horses. *J. Equine Vet. Sci.* **2018**, *67*, 44–49. [CrossRef]
19. Ainsworth, D.M.; Grünig, G.; Matychak, M.B.; Young, J.; Wagner, B.; Erb, H.N.; Antczak, D.F. Recurrent airway obstruction (RAO) in horses is characterized by IFN-γ and IL-8 production in bronchoalveolar lavage cells. *Vet. Immun. Immunopathol.* **2003**, *96*, 83–91. [CrossRef]
20. Venugopal, C.S.; Holmes, E.P.; Polikepahad, S.; Laborde, M.K.; Moore, R.M. Neurokinin receptors in recurrent airway obstruction; a comparative study of affected and unaffected horses. *Can. J. Vet. Res.* **2009**, *73*, 25–33.
21. Buechner-Maxwell, V. Airway hyperresponsiveness. *Compend. Contin. Educ. Pract. Vet.* **1993**, *15*, 1379–1389.
22. Brazil, T.J.; McGorum, B.C. Molecules and inflammation in equine heaves: Mechanism and markers of disease. *Equine Vet. J.* **2001**, *33*, 113–115. [CrossRef]
23. Calzetta, L.; Rogliani, P.; Pistocchini, E.; Mattei, M.; Cito, G.; Alfonsi, P.; Page, C.; Matera, M.G. Effect of lipopolysaccharide on the responsiveness of equine bronchial tissue. *Pulm. Pharmacol. Ther.* **2018**, *49*, 88–94. [CrossRef]
24. Wada, R.; Aida, H.; Kaneko, M.; Oikawa, M.; Yoshihara, T.; Tomioka, Y.; Nitta, M. Identification of the bronchi for bronchoscopy in the horse and segmentation of the horse lung. *Jpn. J. Equine Sci.* **1992**, *3*, 37–43. [CrossRef]
25. Rush, B.R.; Raub, E.S.; Rhoads, W.S.; Flaminio, M.J.; Matson, C.J.; Hakala, J.E.; Gillespie, J.R. Pulmonary function in horses with recurrent airway obstruction after aerosol and parenteral administration of beclomethasone dipropionate and dexamethasone, respectively. *Am. J. Vet. Res.* **1998**, *59*, 1039–1043.
26. Leclere, M.; Lavoie-Lamoureux, A.; Lavoie, J.P. Heaves, an asthma-like disease of horses. *Respirology* **2011**, *16*, 1027–1046. [CrossRef]
27. Ferrari, C.R.; Cooley, J.; Mujahid, N.; Costa, L.R.; Wills, R.W.; Johnson, M.E.; Swiderski, C.E. Horses with pasture asthma have airway remodelling that is characteristic of human asthma. *Vet. Pathol.* **2018**, *55*, 144–158. [CrossRef]
28. Bullone, M.; Lavoie, J.P. Asthma "of horses and men"—How can equine heaves help us better understand human asthma immunopathology and its functional consequences? *Mol. Immunol.* **2015**, *66*, 97–105. [CrossRef]
29. Woort, F.T.; Caswell, J.L.; Arroyo, L.G.; Viel, L. Histologic investigation of airway inflammation in postmortem lung samples from racehorses. *Am. J. Vet. Res.* **2018**, *79*, 342–347. [CrossRef]
30. Giguère, S.; Prescott, J.F. Quantitation of equine cytokine mRNA expression by transcription-competitive polymerase chain reaction. *Vet. Immunol. Immunopathol.* **1999**, *67*, 1–15. [CrossRef]
31. Swiderski, C.E.; Klei, T.R.; Horohov, D.W. Quantitative measurement of equine cytokine mRNA expression by polymerase chain reaction using target-specific standard curves. *J. Immunol. Methods* **1999**, *222*, 155–169. [CrossRef]
32. Joubert, P.; Silversides, D.W.; Lavoie, J.P. Equine neutrophils express mRNA for tumor necrosis factor-α, interleukin (IL)-1β, IL-6, IL-8, macrophage-inflammatory-protein-2 but not for IL-4, IL-5 and interferon-γ. *Equine Vet. J.* **2001**, *33*, 730–733. [CrossRef] [PubMed]
33. Beadle, R.E.; Horohov, D.W.; Gaunt, S.D. Interleukin-4 and interferon-gamma gene expression in summer pasture-associated obstructive pulmonary disease affected horses. *Equine Vet. J.* **2002**, *34*, 389–394. [CrossRef] [PubMed]
34. Joubert, P.; Cordeau, M.E.; Boyer, A.; Silversides, D.W.; Lavoie, J.P. Quantification of mRNA expression by peripheral neutrophils in an animal model of asthma. *Am. J. Respir. Crit. Care Med.* **2002**, *165*, A317.
35. Horohov, D.W.; Beadle, R.E.; Mouch, S.; Pourciau, S.S. Temporal regulation of cytokine mRNA expression in equine recurrent airway obstruction. *Vet. Immunol. Immunopathol.* **2005**, *108*, 237–245. [CrossRef]
36. Laan, T.T.; Bull, S.; Pirie, R.S.; Fink-Gremmels, J. Evaluation of cytokine production by equine alveolar macrophages exposed to lipopolysaccharide, *Aspergillus fumigatus*, and a suspension of hay dust. *Am. J. Vet. Res.* **2005**, *66*, 1584–1589. [CrossRef]
37. Ainsworth, D.M.; Wagner, B.; Franchini, M.; Grünig, G.; Erb, H.N.; Tan, J.Y. Time-dependent alterations in gene expression of interleukin-8 in the bronchial epithelium of horses with recurrent airway obstruction. *Am. J. Vet. Res.* **2006**, *67*, 669–677. [CrossRef]
38. Riihimäki, M.; Raine, A.; Art, T.; Lekeux, P.; Couëtil, L.; Pringle, J. Partial divergence of cytokine mRNA expression in bronchial tissues compared to bronchoalveolar lavage cells in horses with recurrent airway obstruction. *Vet. Immunol. Immunopathol.* **2008**, *122*, 256–264. [CrossRef]
39. Klukowska-Rotzler, J.; Swinburne, J.E.; Drogemuller, C.; Dolf, G.; Janda, J.; Leeb, T.; Gerber, V. The interleukin 4 receptor gene and its role in recurrent airway obstruction in Swiss Warmblood horses. *Anim. Genet.* **2012**, *43*, 450–453. [CrossRef]
40. Padoan, E.; Ferraresso, S.; Pegolo, S.; Castagnaro, M.; Barnini, C.; Bargelloni, L. Real time RT-PCR analysis of inflammatory mediator expression in recurrent airway obstruction-affected horses. *Vet. Immunol. Immunopathol.* **2013**, *156*, 190–199. [CrossRef]
41. Barton, A.K.; Gehlen, G. Pulmonary Remodelling in Equine Asthma: What Do We Know about Mediators of Inflammation in the Horse? *Mediat. Inflamm.* **2016**, *2016*, 5693205. [CrossRef]
42. Kraneveld, A.D.; Nijkamp, F.P.; Van Oosterhout, A.J. Role for neurokinin-2 receptor in interleukin-5-induced airway hyperresponsiveness but not eosinophilia in guinea pigs. *Am. J. Respir. Crit. Care Med.* **1997**, *156*, 367–374. [CrossRef]

43. Fattori, D.; Altamura, M.; Maggi, C.A. Small molecule antagonists of the tachykinin NK 2 receptor. *Mini-Rev. Med. Chem.* **2004**, *4*, 331–340. [CrossRef]
44. Pennefather, J.N.; Lecci, A.; Candenas, M.L.; Patak, E.; Pinto, F.M.; Maggi, C.A. Tachykinins and tachykinin receptors: A growing family. *Life Sci.* **2004**, *74*, 1445–1463. [CrossRef]
45. Maggi, C.A.; Patacchini, R.; Santicioli, P.; Giuliani, S. Tachykinin antagonists and capsaicin-induced contraction of the rat isolated urinary bladder: Evidence for tachykinin-mediated cotransmission. *Br. J. Pharmacol.* **1991**, *103*, 1535–1541. [CrossRef]
46. Krishnakumar, S.; Holmes, E.P.; Moore, R.M.; Kappel, L.; Venugopal, C.S. Non-adrenergic non-cholinergic excitatory innervation in the airways: Role of neurokinin-2 receptors. *Auton. Autacoid Pharmacol.* **2002**, *22*, 215–224. [CrossRef]
47. Schelfhout, V.; Van De Velde, V.; Maggi, C.; Pauwels, R.; Joo, G. The effect of the tachykinin NK2 receptor antagonist MEN11420 (nepadutant) on neurokinin A induced bronchoconstriction in asthmatics. *Ther. Adv. Respir. Dis.* **2009**, *3*, 219–226. [CrossRef]
48. Calzetta, L.; Luongo, L.; Cazzola, M.; Page, C.; Rogliani, P.; Facciolo, F.; Maione, S.; Capuano, A.; Rinaldi, B.; Matera, M.G. Contribution of sensory nerves to LPS-induced hyperresponsiveness of human isolated bronchi. *Life Sci.* **2015**, *131*, 44–50. [CrossRef]
49. Advenier, C.; Lagente, V.; Boichot, E. The role of tachykinin receptor antagonists in the prevention of bronchial hyperresponsiveness, airway inflammation and cough. *Eur. Respir. J.* **1997**, *10*, 1892–1906. [CrossRef]
50. Solinger, N.; Sonea, I.M. Distribution of the neurokinin-1 receptor in equine intestinal smooth muscle. *Equine Vet. J.* **2008**, *40*, 321–325. [CrossRef]
51. Lavoie, J.P.; Maghni, K.; Desnoyers, M.; Rame, T.; Martin, J.G.; Hamid, Q.A. Neutrophilic airway inflammation in horses with heaves is characterized by a Th2-type cytokine profile. *Am. J. Respir. Crit. Care Med.* **2001**, *164*, 1410–1413. [CrossRef]
52. Giguere, S.; Viel, L.; Lee, E.; MacKay, R.J.; Hernandez, J.; Franchini, M. Cytokine induction in pulmonary airways of horses with heaves and effect of therapy with inhaled fluticasone propionate. *Vet. Immunol. Immunopathol.* **2002**, *85*, 147–158. [CrossRef]
53. Cordeau, M.E.; Joubert, P.; Dewachi, O.; Hamid, Q.; Lavoie, J.P. IL-4, IL-5 and IFNgamma mRNA expression in pulmonary lymphocytes in equine heaves. *Vet. Immunol. Immunopathol.* **2004**, *97*, 87–96. [CrossRef]
54. Hansen, S.; Otten, N.D.; Bircha, K.; Skovgaard, K.; Hopster-Iversen, C.; Fjeldborg, J. Bronchoalveolar lavage fluid cytokine, cytology and IgE allergen in horses with equine asthma. *Vet. Immunol. Immunopathol.* **2020**, *220*, 109976. [CrossRef]
55. Moran, G.; Folch, H. Recurrent airway obstruction in horses—An allergic inflammation: A review. *Vet. Med.* **2011**, *56*, 1–13. [CrossRef]
56. Leguillette, R. Recurrent airway obstruction—Heaves. *Vet. Clin. Equine* **2003**, *19*, 63–86. [CrossRef]
57. Laan, T.T.J.M.; Bull, S.; Pirie, R.S.; Fink-Gremmels, J. The role of alveolar macrophages in the pathogenesis of recurrent airway obstruction in horses. *J. Vet. Intern. Med.* **2006**, *20*, 167–174. [CrossRef]
58. Franchini, M.; Gill, U.; Von Fellenberg, R.; Bracher, V.D. Interleukin-8 concentration and neutrophil chemotactic activity in bronchoalveolar lavage fluid of horses with chronic obstructive pulmonary disease following exposure to hay. *Am. J. Vet. Res.* **2000**, *61*, 1369–1374. [CrossRef]
59. Ainsworth, D.M.; Matychak, M.; Reyner, C.L.; Erb, H.N.; Young, J.C. Effects of in vitro exposure to hay dust on the gene expression of chemokines and cell-surface receptors in primary bronchial epithelial cell cultures established from horses with chronic recurrent airway obstruction. *Am. J. Vet. Res.* **2009**, *70*, 365–372. [CrossRef]
60. Ainsworth, D.M.; Wagner, B.; Erb, H.N.; Young, J.C.; Retallick, D.E. Effects of in vitro exposure to hay dust on expression of interleukin-17, -23, -8, and -1beta and chemokine (C-X-C motif) ligand 2 by pulmonary mononuclear cells isolated from horses chronically affected with recurrent airway disease. *Am. J. Vet. Res.* **2007**, *68*, 1361–1369. [CrossRef] [PubMed]
61. Berndt, A.; Derksen, F.J.; Venta, P.J.; Ewart, S.; Yuzbasiyan-Gurkan, V.; Robinson, N.E. Elevated amount of Toll-like receptor 4 mRNA in bronchial epithelial cells is associated with airway inflammation in horses with recurrent airway obstruction. *Am. J. Physiol. Lung Cell. Mol. Physiol.* **2007**, *292*, L936–L943. [CrossRef] [PubMed]
62. Reyner, C.L.; Wagner, B.; Young, J.C.; Ainsworth, D.M. Effects of in vitro exposure to hay dust on expression of interleukin-23, -17, -8, and -1beta and chemokine (C-X-C motif) ligand 2 by pulmonary mononuclear cells from horses susceptible to recurrent airway obstruction. *Am. J. Vet. Res.* **2009**, *70*, 1277–1283. [CrossRef] [PubMed]
63. Pietra, M.; Peli, A.; Bonato, A.; Ducci, A.; Cinotti, S. Equine bronchoalveolar lavage cytokines in the development of Recurrent Airway Obstruction. *Vet. Res. Commun.* **2007**, *31*, 313–316. [CrossRef] [PubMed]
64. Pietra, M.; Cinotti, S.; Ducci, A.; Giunti, M.; Peli, A. Time-dependent changes of cytokines mRNA in bronchoalveolar lavage fluid from symptomatic recurrent airway obstruction-affected horses. *Pol. J. Vet. Sci.* **2011**, *14*, 343–351. [CrossRef]
65. Hansen, S.; Baptiste, K.E.; Fjeldborg, J.; Betancourt, A.; Horohov, D.W. A comparison of pro-inflammatory cytokine mRNA expression in equine bronchoalveolar lavage (BAL) and peripheral blood. *Vet. Immunol. Immunopathol.* **2014**, *158*, 238–243. [CrossRef]

Article

Phenotypic Characterization of Encephalitis and Immune Response in the Brains of Lambs Experimentally Infected with Spanish Goat Encephalitis Virus

Ileana Z. Martínez [1,2], Claudia Pérez-Martínez [1], Luis M. Salinas [1,3], Ramón A. Juste [4], Juan F. García Marín [1,5] and Ana Balseiro [1,5,*]

[1] Departamento de Sanidad Animal, Facultad de Veterinaria, Universidad de León, 24006 León, Spain; ileanazorhaya.martinez@upaep.mx (I.Z.M.); cperm@unileon.es (C.P.-M.); lsalir00@estudiantes.unileon.es (L.M.S.); jfgarm@unileon.es (J.F.G.M.)

[2] Universidad Popular Autónoma del Estado de Puebla, UPAEP Universidad, Puebla 72410, Mexico

[3] Universidad Internacional Antonio de Valdivieso, UNIAV, 47000 Rivas, Nicaragua

[4] Animal Health Department, NEIKER-Instituto Vasco de Investigación y Desarrollo Agrario, 48160 Derio, Bizkaia, Spain; rjuste@neiker.eus

[5] Departamento de Sanidad Animal, Instituto de Ganadería de Montaña, CSIC-Universidad de León, Finca Marzanas, Grulleros, 24346 León, Spain

* Correspondence: abalm@unileon.es

Received: 10 July 2020; Accepted: 5 August 2020; Published: 7 August 2020

Simple Summary: This article studies the local immune response in the central nervous system (CNS) in lambs experimentally infected with Spanish goat encephalitis virus. CNS sections were immunostained to detect microglia, astrocytes, T lymphocytes, and B lymphocytes. In glial foci and perivascular cuffing areas, microglia were the most abundant cell type (45.4% of immunostained cells), followed by T lymphocytes (18.6%) and B lymphocytes (4.4%). Reactive astrogliosis occurred to a greater extent in the lumbosacral spinal cord. Thalamus, hypothalamus, corpus callosum, and medulla oblongata cord contained the largest areas occupied by glial foci. Lesions were more severe in lambs than in goats.

Abstract: Spanish goat encephalitis virus (SGEV), a novel subtype of tick-borne flavivirus closely related to louping ill virus, causes a neurological disease in experimentally infected goats and lambs. Here, the distribution of microglia, T and B lymphocytes, and astrocytes was determined in the encephalon and spinal cord of eight Assaf lambs subcutaneously infected with SGEV. Cells were identified based on immunohistochemical staining against Iba1 (microglia), CD3 (T lymphocytes), CD20 (B lymphocytes), and glial fibrillary acidic protein (astrocytes). In glial foci and perivascular cuffing areas, microglia were the most abundant cell type (45.4% of immunostained cells), followed by T lymphocytes (18.6%) and B lymphocytes (4.4%). Thalamus, hypothalamus, corpus callosum, and medulla oblongata contained the largest areas occupied by glial foci. Reactive astrogliosis occurred to a greater extent in the lumbosacral spinal cord than in other regions of the central nervous system. Lesions were more frequent on the side of the animal experimentally infected with the virus. Lesions were more severe in lambs than in goats, suggesting that lambs may be more susceptible to SGEV, which may be due to species differences or to interindividual differences in the immune response, rather than to differences in the relative proportions of immune cells. Larger studies that monitor natural or experimental infections may help clarify local immune responses to this flavivirus subtype in the central nervous system.

Keywords: Spanish goat encephalitis virus (SGEV); goat; lambs; cell population; immunohistochemistry

---

1. Introduction

Louping ill is a vector-borne disease endemic of the British Isles and Ireland, which is caused by the louping ill virus (LIV) [1]. In 2011, the Spanish goat encephalitis virus (SGEV), a tick-borne flavivirus subtype closely related to LIV [1], naturally induced non-purulent encephalomyelitis in goats in Spain [2]. SGEV also induced clinical signs (febrile illness and neurological signs such as muscular tremors—mainly located in the neck—ataxia, and/or incoordination) and histopathological lesions in the nervous system in experimentally infected goats and lambs [3,4]. In those studies, lambs developed more severe histological lesions in the central nervous system (CNS) than goats, suggesting greater susceptibility to SGEV [4]. This greater susceptibility has been attributed to species differences and interindividual differences in the immune response; in fact, an effective specific inflammatory response against LIV has been shown to neutralize the spread of the infection in the nervous system in lambs [5].

The pathogenesis of an acute viral infection by flavivirus in the CNS involves complex virus–host interactions in which the recruitment of immune cells into the CNS plays a fundamental role in the outcome of the disease. The response to SGEV in goat CNS has been shown to comprise microglia, T lymphocytes, and, a to lesser extent, B lymphocytes [6]. This is similar to the responses to LIV in mice and lambs [7], to West Nile virus in humans and horses [8,9], to Japanese encephalitis virus in humans [10], and to tick-borne encephalitis virus in humans and non-human primates [11,12].

A detailed understanding of the immune response to SGEV in the CNS of different animal species may help clarify the differences among flavivirus. Therefore, the present study examined the phenotype and distribution of microglia, T and B lymphocytes, and astrocytes in the CNS of lambs experimentally infected with SGEV, and the results were compared with the previously reported immune response in goats [6].

2. Materials and Methods

2.1. Animals and Sampling

Samples were obtained from eight female 3-month-old Assaf lambs (identified as 14, 18, 19, 21, 25, 26, 28, and 29) from a previous study [4]. In that work, the animals were challenged subcutaneously on the right thorax caudal to the elbow with 1 mL of a suspension containing 1.0×10^7 plaque-forming units per mL of SGEV grown in BHK-21 baby hamster kidney cells. The following postmortem CNS samples were analyzed by histopathology in that study: cerebral cortex, corpus callosum, thalamus, hypothalamus, hippocampus, midbrain, cerebellum, pons, medulla oblongata, and four sections of spinal cord (cervical, thoracic, lumbar, and sacral). Samples were placed in 10% neutral buffered formalin, processed routinely through graded alcohols, and embedded in paraffin wax. The severity of microscopic lesions was rated using our previously reported scale [4]: grade I, only perivascular cuffing; grade II, perivascular cuffing and small foci of glial cells; grade III, moderate non-suppurative encephalomyelitis, perivascular cuffing, diffuse or focal proliferation of glial cells, neuronal degeneration, neuron necrosis, and neuronophagia, demyelination, vacuolation of the neuropil, meningitis, and microvascular changes consisting of reactive endothelium and perivascular edema; and grade IV, which involved the same characteristics as grade III but more severe. In the eight animals, lesions were more frequent and severe in the midbrain, cerebellum, medulla oblongata, and all sections of the spinal cord than in other tissues examined [4].

2.2. Immunohistochemistry

Serial paraffin-embedded sections (3 μm) were prepared from the CNS samples listed in Section 2.1. and stained with primary antibodies against the following antigens: ionized calcium-binding adaptor molecule 1 (Iba1) to detect microglia cells [9], CD3 to detect T lymphocytes, CD20 to detect B lymphocytes, and glial fibrillary acidic protein (GFAP) to detect astrocytes (Table 1). The sections were de-paraffinized, and endogenous peroxidase activity was blocked by incubating them with 0.5% H_2O_2 in distilled water for 30 min. Following antigen retrieval (Table 1), the samples were incubated for 20 min in a humidified chamber with 5% normal serum and 0.1% bovine serum albumin in Tris-buffered saline to prevent non-specific binding. The remaining steps in the protocol and the reagents used throughout immunohistochemistry staining were identical to those described by Martínez et al. (2020) [6]. As a negative control, the slides were subjected to the standard procedure in the absence of the primary antibodies. As a positive control, lamb lymph node tissue was stained against Iba1, CD3, and CD20. As an overall negative control, samples of CNS from healthy lambs were stained with anti-GFAP antibody, and, as positive control, samples from a sheep with CNS degenerative chronic disease were used.

Table 1. Immunohistochemical protocols used for cellular type characterization.

Antibody/Specificity	Clone n°	Isotype	Manufacturer	Antigen Retrieval	Dilution
Iba1/Macrophages/ microgial cells	Polyclonal 019-19741	Rabbit IgG	FLUJIFILM-Wako Chemicals Europe GmbH, Neuss, Germany	Citrate pH 6.0 in microwave in 20 min	1:1000
CD3/T lymphocytes	Monoclonal NCL-L-CD3-565	Mouse IgG	Novocastra, Leica Biosystem, Neucastle, United Kingdom	Citrate pH 6.0 in microwave in 20 min	1:500
CD20/B lymphocytes	Polyclonal PA5-16701	Rabbit IgG	ThermoFisher, Massachusetts, USA.	Citrate pH 6.0 in steamer for 20 min	1:200
GFAP/Astrocytes	Monoclonal MCA-5C10	Mouse IgG	EncorBiotechnology, Gainesville, Florida, USA	Citrate pH 6.0 in microwave in 20 min	1:8000

GFAP, glial fibrillary acidic protein; Iba1, ionized calcium binding adaptor molecule 1.

2.3. Evaluation and Quantitation

The stained slides were analyzed under a light microscope (Eclipse E600, Nikon, Japan) equipped with a digital camera (DS-Fi1, Nikon). Cells were counted using NIS-Elements BR imaging software (Nikon). Sections were examined in five randomly selected parenchymal fields at a magnification of 200× [9]. The abundance of microglia, T lymphocytes, and B lymphocytes was expressed as a proportion (%) of all immunolabeled cells within the perivascular cuffing and glial foci. Astrocytes were evaluated using the following semi-quantitative scoring system adapted from a previous study on reactive astrogliosis in human disorders [13]: 1, astrocytes present in healthy CNS tissue, some astrocytes do not express detectable levels of GFAP; 2, mild to moderate reactive astrogliosis; 3, severe diffuse reactive astrogliosis; and 4, severe reactive astrogliosis with compact glial scar formation.

In order to estimate the total area occupied by glial foci, the areas with glial foci were summed and divided by the total area of the CNS section [6] using Image J software (https://imagej.nih.gov/ij/download.html). To evaluate the spatial distribution of glial foci, we chose medulla oblongata sections because they were consistently affected and had a similar size in all lambs examined [4] and we studied them using light microscopy (BX51, Olympus, Japan) and the imaging software OlyVIA 2.91 (Olympus, Japan). The images were analyzed using VS-ASW 2.8 software (Olympus, Japan).

2.4. Statistical Analysis

The proportions of immunolabeled cells that were microglia or T or B lymphocytes were submitted to an arcsin square-root transformation in order to meet the normality criterion for statistical analyses. Then, the results were subjected to cluster analysis by the centroid method in the SAS CLUSTER procedure (SAS, Cary, NC, USA), which generated two clusters: cluster A comprised lambs 19, 21, 25, and 29, which showed a relatively loose relationship; and cluster B comprised lambs 14, 18, 26, and 28, which showed a relatively tight relationship. Microscopic lesions in each cluster did not fit nearly within the four-grade scoring system: cluster A lesions were mild to moderate (grades I and III), while cluster B lesions were moderate to severe (Figure 1) (grades III and IV) [4]. These clusters were considered to better represent the severity of lesions in the study sample of lambs. As a consequence, "cluster" was added as a variable in order to capture population variability potentially more accurately than only with individual values.

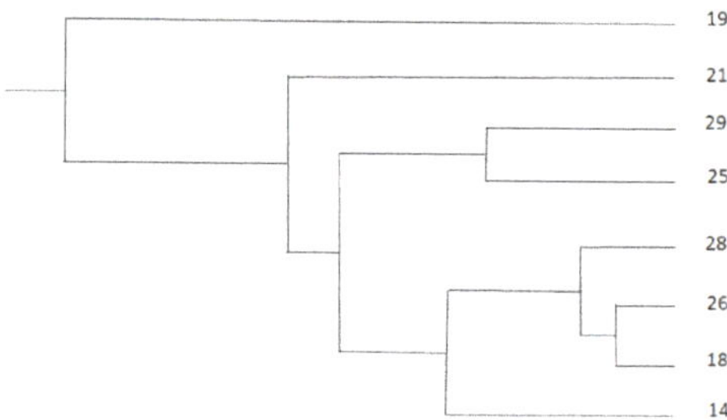

Figure 1. Cluster analysis of lambs experimentally infected with Spanish goat encephalitis virus (SGEV) in terms of lesion severity using the centroid method in the SAS CLUSTER procedure. Lambs 19, 21, 25, and 29 were assigned to cluster A (mild to moderate lesions), while lambs 14, 18, 26, and 28 were assigned to cluster B (moderate to severe lesions).

A general linear model was constructed using the GLM procedure in SAS with the following dependent variables: arcsin-transformed proportions of microglia, T lymphocytes, and B lymphocytes; log-transformed semi-quantitative scores for astrocytes; and proportion of total stained area that was positive for Iba1. The independent variables were: CNS region, type of lesion (perivascular cuffing or glial foci), cluster, and their first-order interactions.

Pairwise least-square means comparisons were carried out using Student's t test as implemented in the LSMEANS statement in SAS. Statistical significance was accepted at $p < 0.05$, while $p = 0.05–0.1$ was taken to indicate a tendency [6].

Potential differences in the number of lesions (glial foci) between the infected side of the animal and the opposite side (laterality effects) were examined using the chi-squared test in the FREQ procedure in SAS and confirmed using the log-transformed foci counts submitted to analysis of variance of the generalized linear model. Laterality effects were also confirmed using a least-squares mean Student's t test with the LSMEANS statement in a model with three independent variables: cluster, side, and their interaction [6].

2.5. Ethics Approval

Sampling procedures and SGEV challenge in the previous study by Salinas et al. (2017) [4] were approved by the Animal Research Ethics Committee of the Community of Junta de Castilla y León, Spain (reference number ULE_010_2015). Experiments were conducted in accordance with Spanish and European legal requirements and guidelines on animal experimentation and welfare.

3. Results

3.1. Microglia

Microglia was the most abundant cellular type (Figure 2a,d), accounting for an average of 45.4% of the three immunostained cell populations. Based on lesion severity, lambs were classified into cluster A with mild to moderate lesions and cluster B with moderate to severe lesions (Section 2.4). Microglia was significantly more abundant in cluster B (59.3%) than in cluster A (31.5%, $p < 0.0001$; Table 2).

Microglia was also significantly more abundant in cortex (55.3%) and hippocampus (55.2%) than in cervical spinal cord (34.4%) or lumbosacral spinal cord (31.6%; both $p < 0.0001$). Additionally, they were significantly more abundant in glial foci (55.2%) than in perivascular cuffing (35.5%, $p < 0.0001$).

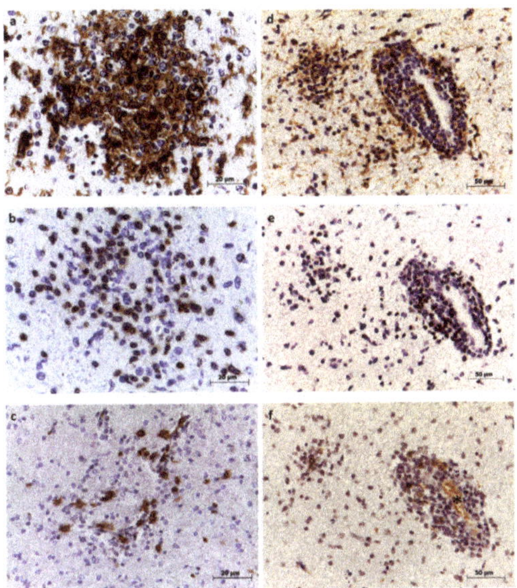

Figure 2. Immunohistochemistry to detect microglia, T lymphocytes, and B lymphocytes in thalamus (**a–c**) and hippocampus (**d–f**). Microglia were identified based on labeling with anti-Iba1 antibodies. Abundant cells are visible within glial foci (**a**) and in the perivascular infiltrate (**d**). T lymphocytes were identified based on labeling with anti-CD3 antibodies. Moderate numbers are visible in the glial focus (**b**) and in the perivascular cuff (**e**). B lymphocytes were identified based on labeling with anti-CD20 antibodies. Few numbers are visible within the glial focus (**c**) and perivascular infiltrate (**f**). Erythrocytes were also immunolabelled (asterisk) but were not included in the counting.

Table 2. Proportion (%) of all immunostained cells, i.e., microglia, T or B lymphocytes, or astrocytes, in lambs experimentally infected with SGEV, stratified according to lesion severity, central nervous system (CNS) region, and lesion type.

Variable	Level	Microglia (Iba1)		T lymphocytes (CD3)		B lymphocytes (CD20)		Astrocytes (GFAP)	
		Mean	SE	Mean	SE	Mean	SE	Mean	SE
Cluster	A (mild to moderate lesions)	31.5	2.3	12.5	2.3	2.4	2.3	2.4	1.1
	B (moderate to severe lesion)	59.3	2.3	24.8	2.34	6.4	2.34	2.3	1.1
Region	Cortex	55.3	4.7	16.4	4.7	3.4	4.70	2.1	2.2
	Thalamus, hypothalamus, corpus callosum	51.4	4.7	18.9	4.7	4.8	4.7	2.4	2.2
	Hippocampus	55.2	4.7	16.0	4.7	3.9	4.7	2.0	2.2
	Midbrain	42.1	4.7	21.6	4.7	5.8	4.7	2.1	2.2
	Pons, cerebellum	45.8	4.7	25.1	4.7	4.5	4.7	2.3	2.2
	Medulla oblongata	47.6	4.7	24.7	4.7	3.4	4.7	2.5	2.2
	Cervical spinal cord	34.4	4.7	11.6	4.7	5.6	4.7	2.5	2.2
	Lumbosacral spinal cord	31.6	4.7	14.8	4.7	3.8	4.7	2.9	2.2
Type of lesion	Glial foci	55.3	2.3	16.6	2.3	2.5	2.3	2.3	0.1
	Perivascular Cuffing	35.5	2.3	20.6	2.3	6.3	2.3	-	-

SE, standard error. *Between clusters:* Proportions of microglia and T lymphocytes were significantly different ($p < 0.0001$), whereas proportions of B lymphocytes ($p = 0.1213$) and astrocytes ($p = 0.3435$) were not. *Among CNS regions:* Proportions of microglia in cortex and hippocampus were different from proportions in cervical and lumbosacral spinal cord ($p < 0.0001$). Proportions of T lymphocytes in pons and cerebellum differed from those in cervical spinal cord ($p = 0.0139$). Proportions of B lymphocytes did not differ significantly among regions ($p = 0.5056$), nor did proportions of astrocytes ($p = 0.2124$). *Between type of lesion:* Proportions of microglia differed ($p < 0.0001$), whereas proportions of T lymphocytes ($p = 0.8350$) or B lymphocytes ($p = 0.8659$) did not.

Glial foci were significantly more numerous in cluster B, where they covered a significantly larger proportion of affected areas (15.9%) than in cluster A (8.4%, $p < 0.0001$). Glial foci were significantly more numerous in thalamus, hypothalamus, corpus callosum, and medulla oblongata (19.5% of affected areas) than in cervical spinal cord (5.6%, $p = 0.0682$) and lumbosacral spinal cord (6%, $p = 0.0522$).

3.2. T lymphocytes

T lymphocytes were the second most abundant cell type (Figure 2b,e), accounting for 18.6% of the three immunostained cell populations. Like microglia, they were significantly more abundant in cluster B (24.8%) than in cluster A (12.5%; $p < 0.0001$; Table 2). Their proportion was higher in pons and cerebellum (25.1%) than in cervical spinal cord (11.6%, $p = 0.0139$) and was not significantly different between perivascular cuffing (20.6%) and glial foci (16.6%, $p = 0.8350$).

3.3. B lymphocytes

B lymphocytes were the less abundant cell type (4.4%) (Figure 2c,f), and their proportion was similar in cluster B (6.4%) and cluster A (2.4%, $p = 0.1213$; Table 2). All CNS regions contained comparable proportions of B lymphocytes, ranging from 3.4% in cortex and medulla oblongata to 5.8% in midbrain. Their proportions were slightly higher but not significant in perivascular cuffing (6.3%) than in glial foci (2.5%, $p = 0.8659$).

3.4. Proportions of Microglia, T lymphocyte,s and B Lymphocytes According to Lesion Location and Severity

The proportions of microglia were higher than those of T and B lymphocytes in all CNS regions (Table 2), and the differences were significant in the cortex (microglia vs. B lymphocytes, $p < 0.0001$; microglia vs. T lymphocytes, $p = 0.0051$), thalamus, hypothalamus, and corpus callosum (microglia vs. B lymphocytes, $p < 0.0001$), hippocampus (microglia vs. B lymphocytes, $p < 0.0001$; microglia vs. T lymphocytes, $p = 0.0051$), midbrain (microglia vs. B lymphocytes, $p = 0.0047$), pons and cerebellum (microglia vs. B lymphocytes, $p = 0.0002$), and medulla oblongata (microglia vs. B lymphocytes, $p = 0.0001$). Proportions of T and B lymphocytes were not significantly different in any region.

The proportion of microglia cells was higher than those of both lymphocyte types in cluster B (both $p < 0.0001$), and the proportion of T lymphocytes was higher than that of B lymphocytes ($p < 0.0001$). Similar results were observed in cluster A. The proportions of microglia were not significantly different between clusters A and B (31.5% vs. 59.3%; $p = 0.1213$), but the proportions of lymphocytes were significantly higher in cluster B (T, 24.8% vs. 12.5%, $p < 0.0001$; B, 6.4% vs. 2.4%, $p < 0.0001$).

3.5. Astrocytes

Astrocytes were scored on a four-point scale (Figure 3), and we observed no cases with score 4. The frequencies of astrocyte scores did not differ significantly between regions or clusters. The mean astrocyte score was the highest in the lumbosacral spinal cord (2.9%), decreased in the cervical spinal cord and medulla oblongata (2.5%), and was the lowest in the hippocampus (2%, Table 2).

3.6. Lesion Count and Laterality

Lesions were marginally more frequent on the right side of the CNS (62.3%) (Figure 4), the side of virus injection, than on the left side (37.7%, $p = 0.0450$), although the total mean lesion count did not significantly differ between the right and the left sides (11.9 vs. 6.79, $p = 0.3242$). Similarly, differences in lesion numbers between the right and the left sides were not significant in cluster A (1.3 vs. 0.0, $p = 0.2886$) or cluster B (4.6 vs. 2.0, $p = 0.224$).

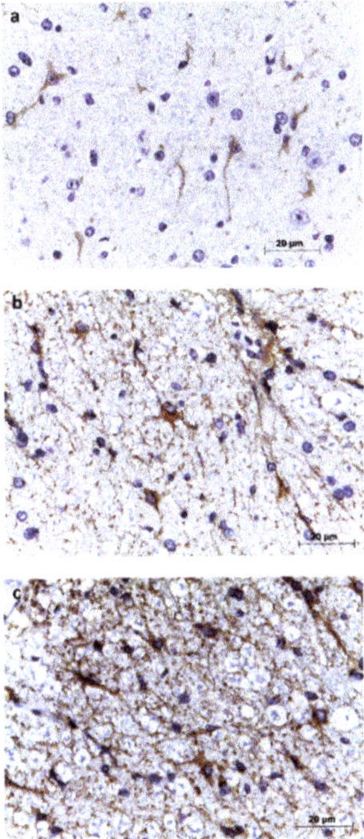

Figure 3. Grading of reactive astrogliosis in cervical spinal cord (**a–c**) using antibodies against GFAP. (**a**) Representative micrograph showing grade 1 (low) reactive astrogliosis. (**b**) Representative micrograph showing grade 2 (mild to moderate) reactive astrogliosis. (**c**) Representative micrograph showing grade 3 (severe) reactive astrogliosis.

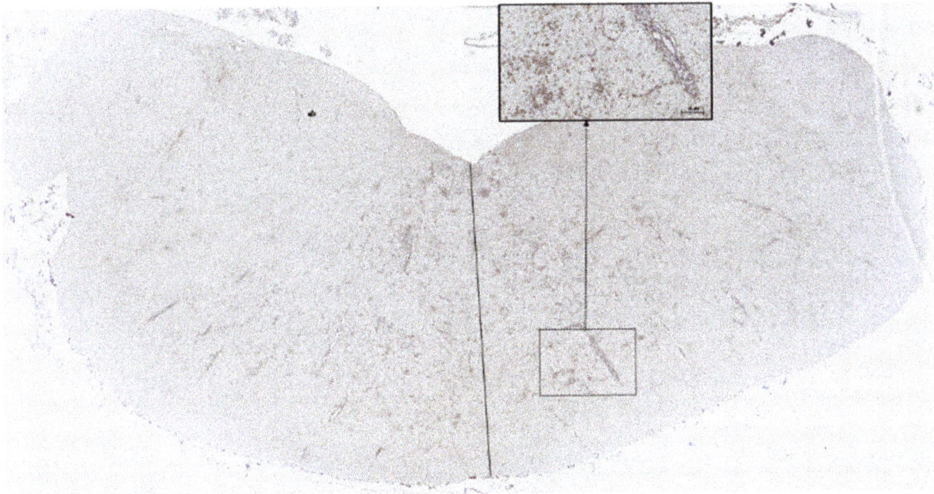

Figure 4. Complete cross-section of the medulla oblongata immunostained for microglia, showing numerous glial foci on the same side of the tissue (right side) as virus injection. Inset: detail of the affected parenchyma.

4. Discussion

This study characterized and quantified the distribution of immune cells in the CNS of lambs experimentally infected with SGEV. The inflammatory cells were predominantly microglia, with a moderate number of T lymphocytes and a smaller number of B lymphocytes. These findings are similar to those of previous studies on SGEV in goats [6] and on LIV and other flaviviruses in horses, humans, and non-human primates [7–12,14,15]. Microglia, which are of mesodermal origin, play a key role in the innate and adaptive immune responses in the CNS [10]; in fact, they are the first cells that respond to CNS infection [12,16]. Infection itself and the resulting neuronal damage activate the recovery in microglia of their amoeboid properties such as motility and of their macrophagic function [10,12,16]. This activation increases neuronal cell death, astrogliosis, and production of pro-inflammatory cytokines [10,12,16].

The proportion of microglia was higher in lambs in the present study than in goats in previous work [6], which may reflect the greater susceptibility of lambs to tissue damage [17]. In this regard, 5 out of 9 and 2 out of 9 lambs showed type III and type IV microscopic lesions, respectively, whilst only 4 out of 9 goats showed grade III lesions, and none grade IV lesions with an identical challenge [3,4]. Moreover, in the present study the proportion of microglia was higher in cluster B and was related to more extensive tissue damage. In studies of human and animal infection with Japanese encephalitis virus, microglia predominated in the thalamus and hippocampus, the primary affected brain regions [18]. We also observed the greatest relative abundance of microglia in thalamus and hippocampus, although these regions were not the most seriously affected by SGEV [4].

T lymphocytes were the second most abundant cell population, as reported in flavivirus infections in goats, lambs, humans, non-human primates, horses, and mice [6,7,9,11,15,17,19]. T lymphocytes control viral infections in the CNS by destroying infected cells, producing cytokines, stimulating phagocytic activity of microglia, and stimulating local antibody production by B lymphocytes [12,20]. High numbers of T lymphocytes have been linked to fatal outcome in human patients infected with tick-borne encephalitis virus [21]. T lymphocytes were more abundant than B lymphocytes, but the difference was not significant, similar to the case in goats infected with SGEV [6] and in lambs infected with LIV [22]. A higher abundance of T lymphocytes than of B lymphocytes has been associated with acute disease in horses infected with West Nile virus [9].

B lymphocytes were the less numerous cellular type and were more abundant in perivascular spaces (6.3% of all immunostained cells) than in parenchyma (2.5%), similar to what has been reported in goats, humans, and non-human primates infected with flavivirus [6,11,12,17,23]. This might reflect antibody-mediated tissue damage in lambs [17]. The proportion of B lymphocytes was lower in lambs in the present study than in goats in previous work [6], which may reflect a lower specific humoral cell-mediated response in lambs.

Reactive astrogliosis was observed in the CNS of mice, lambs, goats, humans, and mice infected with flaviruses [6,23–25], especially in persistent infections [25]. In our study of lambs, most CNS sections received an astrogliosis score of 2, characteristic of diffuse innate immune activation in response to viral infection [13]. An astrogliosis score of 3 was observed mainly in the spinal cord near the pia mater, which is the most likely entry route for the virus into the CNS, where astrocytes are in the lining of brain capillaries and pia mater and are the first cells to come in contact with the virus [25]. GFAP expression has been associated with the greatest damage in humans infected with West Nile virus [23]; this is similar to the present study's finding of a particularly strong GFAP immunostaining in spinal cord and medulla oblongata, where histological lesions following SGEV infection are more frequent and severe [4].

In our SGEV-infected lambs, glial foci were significantly more abundant in thalamus, hypothalamus, corpus callosum, and medulla oblongata; SGEV-infected goats, in contrast, showed more foci in medulla oblongata and spinal cord [6]. Therefore, immunohistochemistry can help identify tissues preferentially affected by different flaviviruses, otherwise difficult to detect using conventional hematoxylin-and-eosin staining, as already demonstrated for tick-borne encephalitis and Japanese encephalitis [14,16]. We observed lesion predominance on the same side of the medulla oblongata as that of the subcutaneous virus injection, similar to results obtained in goats [6]. This preference supports a neurotropic route of virus movement [4].

5. Conclusions

The immune response to SGEV in lambs appears to involve a combination of microglia and T lymphocytes, suggesting that B lymphocytes proliferate during infection. The proportion of microglia among immune cells was higher in SGEV-infected lambs than in goats, which may reflect the greater lesion severity in sheep. Nevertheless, the relative proportions of four types of cells (microglia, T and B lymphocytes, and astrocytes) in the CNS were similar between infected lambs and goats. Therefore, we may conclude that lambs might be more susceptible to SGEV due to species differences or to different immune responses between individuals, rather than to differences in the relative proportion of immune cells. Larger studies that analyze flavivirus infections over longer periods may further increase our understanding of the immune response.

Author Contributions: Conceptualization, A.B.; methodology, I.Z.M., L.M.S., C.P.-M., J.F.G.M., R.A.J., and A.B.; validation, C.P.-M., J.F.G.M., R.A.J., and A.B.; formal analysis, I.Z.M., C.P.-M., J.F.G.M., R.A.J., and A.B.; investigation, I.Z.M. and A.B.; resources, A.B.; data curation, C.P.-M., J.F.G.M., R.A.J., and A.B.; writing—original draft preparation, I.Z.M. and A.B.; writing—review and editing, I.Z.M., C.P.-M., L.M.S., R.A.J., J.F.G.M., and A.B. All authors have read and agreed to the published version of the manuscript.

Funding: This work was partially supported by a FEDER co-funded grant from the Instituto Nacional de Investigación y Tecnología Agraria y Alimentaria (E-RTA2013-00013-C04-04), Ministerio de Ciencia, Innovación y Universidades (MCIU) and the Agencia Estatal de Investigación (AEI) reference project RTI2018-096010-B-C21 (FEDER co-funded) and by the Principado de Asturias, PCTI 2018–220 (GRUPIN: IDI2018-000237 and FEDER). Ms. Ileana Z. Martínez was supported by a Fundación Carolina PhD scholarship (2017 call).

Acknowledgments: The authors would like to thank the Vice-Ministry of the Environment of the Principado of Asturias. The authors thank A. Chapin Rodríguez for critically reviewing the manuscript. Authors also thank Red de Investigación en Sanidad Animal (RISA, Spain).

Conflicts of Interest: The authors declare no conflict of interest.

References

1. Mansfield, K.L.; Balseiro Morales, A.; Johnson, N.; Ayllón, N.; Höfle, U.; Alberdi, P.; Fernández de Mera, I.G.; García Marín, J.F.; Gortázar, C.; De la Fuente, J.; et al. Identification and characterization of a novel tick-borne flavivirus subtype in goats (*Capra hircus*) in Spain. *J. Gen. Virol.* **2015**, *96*, 1676–1681. [CrossRef] [PubMed]
2. Balseiro, A.; Royo, L.J.; Pérez-Martínez, C.; de Mera, I.G.F.; Höfle, Ú.; Polledo, L.; Marreros, N.; Casais, R.; García-Marín, J.F. Louping ill in goats, Spain, 2011. *Emerg. Infect. Dis.* **2012**, *18*, 976–978. [CrossRef] [PubMed]
3. Salinas, L.M.; Casais, R.; García Marín, J.F.; Dalton, K.P.; Royo, L.J.; Del Cerro, A.; Gayo, E.; Dagleish, M.P.; Alberdi, P.; Juste, R.A.; et al. Vaccination against louping ill virus protects goats from experimental challenge with Spanish goat encephalitis virus. *J. Comp. Pathol.* **2017**, *156*, 409–418. [CrossRef] [PubMed]
4. Salinas, L.M.; Casais, R.; García Marín, J.F.; Dalton, K.P.; Royo, L.J.; del Cerro, A.; Gayo, E.; Dagleish, M.P.; Juste, R.A.; Balseiro, A. Lambs are susceptible to experimental challenge with Spanish goat encephalitis virus. *J. Comp. Pathol.* **2017**, *156*, 400–413. [CrossRef] [PubMed]
5. Doherty, P.C.; Reid, H.E.; Smith, W. Loping—Ill encephalomyelitis in the sheep. IV. Nature perivascular inflammatory reaction. *J. Comp. Pathol.* **1971**, *81*, 545–549. [CrossRef]
6. Martínez, I.Z.; Pérez-Martínez, C.; Salinas, L.M.; García-Marín, J.F.; Juste, R.A.; Balseiro, A. Phenotypic characterization of encephalitis in the brain of goats experimentally infected with Spanish goat encephalitis virus. *Vet. Immnol. Immunopathol.* **2020**, *220*, 1–7. [CrossRef]
7. Sheahan, B.J.; Moore, M.; Atkins, G.J. The pathogenicity of louping ill virus for mice and lambs. *J. Comp. Pathol.* **2002**, *126*, 137–146. [CrossRef]
8. Cho, H.; Diamond, M.S. Immune responses to West Nile virus infection in the central nervous system. *Viruses* **2012**, *4*, 3814–3830. [CrossRef]

9. Delcambre, G.H.; Liu, J.; Streit, W.J.; Shaw, G.P.J.; Vallario, K.; Herrington, J.; Wenzlow, N.; Barr, K.L.; Long, M.T. Phenotypic characterization of cell populations in the brains of horses experimentally infected with West Nile virus. *Equine Vet. J.* **2017**, *49*, 815–820. [CrossRef]

10. Thongtan, T.; Thepparit, C.; Smith, D.R. The involvement of microglia cells in Japanese encephalitis infections. *Clin. Dev. Immunol.* **2012**, *2012*, 1–7. [CrossRef]

11. Maximova, O.A.; Faucette, L.J.; Ward, J.M.; Murphy, B.R.; Pletnev, A.G. Cellular inflammatory response to flaviviruses in the central nervous system of a primate host. *J. Histochem. Cytochem.* **2009**, *57*, 973–989. [CrossRef] [PubMed]

12. Maximova, O.A.; Pletnev, A.G. Flaviviruses and the central nervous system: Revisiting neuropathological concepts. *Annu. Rev. Virol.* **2018**, *5*, 255–272. [CrossRef] [PubMed]

13. Sofroniew, M.V.; Vinters, H.V. Astrocytes: Biology and pathology. *Acta Neuropathol.* **2010**, *119*, 7–35. [CrossRef] [PubMed]

14. Johnson, R.T.; Burke, D.S.; Elwell, M.; Leake, C.J.; Nisalak, A.; Hoke, C.H.; Lorsomrudee, W. Japanese encephalitis: Immunocytochemical studies of viral antigen and inflammatory cells in fatal cases. *Ann. Neurol.* **1985**, *18*, 567–573. [CrossRef] [PubMed]

15. Gelpi, E.; Preusser, M.; Garzuly, F.; Holzmann, H.; Heinz, F.X.; Budka, H. Visualization of central European tick-borne encephalitis infection in fatal human cases. *J. Neuropathol. Exp. Neurol.* **2005**, *64*, 506–512. [CrossRef]

16. Rock, R.B.; Gekker, G.; Hu, S.; Sheng, W.S.; Cheeran, M.; Lokensgard, J.R.; Peterson, P.K. Role of microglia in central nervous system infections. *Clin. Microbiol. Rev.* **2004**, *17*, 942–964. [CrossRef]

17. Gelpi, E.; Preusser, M.; Laggner, U.; Garzuly, F.; Holzmann, H.; Heinz, F.; Budka, H. Inflammatory response in human tick-borne encephalitis: Analysis of postmortem brain tissue. *J. Neurovirol.* **2006**, *12*, 322–327. [CrossRef]

18. Ghoshal, A.; Das, S.; Ghosh, S.; Mishra, M.K.; Sharma, V.; Koli, P.; Sen, E.; Basu, A. Proinflammatory mediators released by activated microglia induces neuronal death in Japanese encephalitis. *Glia* **2007**, *55*, 483–496. [CrossRef]

19. Blom, K.; Cuapio, A.; Sandberg, J.T.; Varnaite, R.; Michaëlsson, J.; Björkström, N.K.; Sandberg, J.K.; Klingström, J.; Lindquist, L.; Gredmark, R.S.; et al. Cell-mediated immune responses and immunopathogenesis of human tick-borne encephalitis virus-infection. *Front. Immunol.* **2018**, *9*, 1–10. [CrossRef]

20. Binder, G.K. Interferon-gamma- mediated site-specific clearance of alphavirus from CNS neurons. *Science* **2001**, *293*, 303–306. [CrossRef]

21. Ruzek, D.; Salat, J.; Palus, M.; Gritsun, T.S.; Gould, E.A.; Dyková, I.; Skallová, A.; Jelínek, J.; Kopecký, J.; Grubhoffer, L. CD8+ T-cells mediate immunopathology in tick-borne encephalitis. *Virology* **2009**, *384*, 1–6. [CrossRef] [PubMed]

22. Mansfield, K.L.; Johnson, N.; Banyard, A.C.; Núñez, A.; Baylis, M.; Solomon, T.; Fooks, A.R. Innate and adaptive immune responses to tick-borne flavivirus infections in sheep. *Vet. Microbiol.* **2016**, *185*, 20–28. [CrossRef] [PubMed]

23. Kelley, T.W.; Prayson, R.A.; Ruiz, A.I.; Isada, C.M.; Gordon, S.M. The neuropathology of West Nile virus meningoencephalitis. A report of two cases and review of the literature. *Am. J. Clin. Pathol.* **2003**, *119*, 749–753. [CrossRef] [PubMed]

24. Palus, M.; Bily, T.; Elsterova, J.; Langhansova, H.; Salat, J.; Vancova, M.; Rek, D. Infection and injury of human astrocytes by tick-borne encephalitis virus. *J. Gen. Virol.* **2014**, *95*, 2411–2426. [CrossRef] [PubMed]

25. Potokar, M.; Jorgačevski, J.; Zorec, R. Astrocytes in flavivirus infections. *Int. J. Mol. Sci.* **2019**, *20*, 691. [CrossRef] [PubMed]

Article

Immunohistochemical Assessment of Immune Response in the Dermis of *Sarcoptes scabiei*—Infested Wild Carnivores (Wolf and Fox) and Ruminants (Chamois and Red Deer)

Ileana Z. Martínez [1,2], Álvaro Oleaga [3], Irene Sojo [1], María José García-Iglesias [1], Claudia Pérez-Martínez [1], Juan F. García Marín [1,4] and Ana Balseiro [1,4,*]

[1]   Departamento de Sanidad Animal, Facultad de Veterinaria, Universidad de León, 24006 León, Spain; ileanazorhaya.martinez@upaep.mx (I.Z.M.); isojoa00@estudiantes.unileon.es (I.S.); mjgari@unileon.es (M.J.G.-I.); cperm@unileon.es (C.P.-M.); jfgarm@unileon.es (J.F.G.M.)
[2]   Universidad Popular Autónoma del Estado de Puebla, UPAEP Universidad, 72410 Puebla, Mexico
[3]   SERPA, Sociedad de Servicios del Principado de Asturias S.A., 33202 Gijón, Spain; alvaroleaga@yahoo.es
[4]   Departamento de Sanidad Animal, Instituto de Ganadería de Montaña, CSIC-Universidad de León, Finca Marzanas, Grulleros, 24346 León, Spain
*   Correspondence: abalm@unileon.es

Received: 3 June 2020; Accepted: 3 July 2020; Published: 6 July 2020

Simple Summary: This article studies the local immune processes in dermis underlying the macroscopical differences (hyperkeratotic or alopecic) in mangy lesions from wolves (*Canis lupus*), foxes (*Vulpes vulpes*), chamois (*Rupicapra rupicapra*) and red deer (*Cervus elaphus*) naturally infested with *Sarcoptes scabiei*. Skin sections were immuno-stained to detect macrophages, plasma cells, T lymphocytes and B lymphocytes. Skin lesions contained significantly more inflammatory cells in fox than in wolf and chamois. Macrophages were the most abundant inflammatory cells in the lesions of all the species studied, suggesting a predominantly innate, non-specific immune response. Lesions from wolf contained higher proportions of macrophages than the other species, which may reflect a more effective response, leading to alopecic lesions. Fox and chamois may also mount substantial humoral and cellular immune responses with apparently scarce effectiveness that lead to hyperkeratotic lesions.

Abstract: Sarcoptic mange is caused by the mite *Sarcoptes scabiei* and has been described in several species of domestic and wild mammals. Macroscopic lesions are predominantly hyperkeratotic (type I hypersensitivity) in fox, chamois and deer, but alopecic (type IV hypersensitivity) in wolf and some fox populations. To begin to understand the immune processes underlying these species differences in lesions, we examined skin biopsies from wolves (*Canis lupus*), foxes (*Vulpes vulpes*), chamois (*Rupicapra rupicapra*) and red deer (*Cervus elaphus*) naturally infested with *S. scabiei*. Twenty skin samples from five animals per species were used. Sections were immuno-stained with primary antibodies against Iba1 to detect macrophages, lambda chain to detect plasma cells, CD3 to detect T lymphocytes and CD20 to detect B lymphocytes. Skin lesions contained significantly more inflammatory cells in the fox than in the wolf and chamois. Macrophages were the most abundant inflammatory cells in the lesions of all the species studied, suggesting a predominantly innate, non-specific immune response. Lesions from the wolf contained higher proportions of macrophages than the other species, which may reflect a more effective response, leading to alopecic lesions. In red deer, macrophages were significantly more abundant than plasma cells, T lymphocytes and B lymphocytes, which were similarly abundant. The fox proportion of plasma cells was significantly higher than those of T and B lymphocytes. In chamois, T lymphocytes were more abundant than B lymphocytes and plasma cells, although the differences were significant only in the case of macrophages. These results suggest that all the species examined mount a predominantly innate immune response against *S. scabiei* infestation, while fox and chamois may also mount substantial humoral and cellular immune responses, respectively, with apparently scarce effectiveness that lead to hyperkeratotic lesions.

Keywords: *Sarcoptes scabiei*; dermis cellular response; wolf; red fox; chamois; red deer; immunohistochemistry

1. Introduction

Sarcoptic mange is a parasitic skin disease that has been described in several species of domestic and wild mammals, and it is also a zoonosis that affects humans. It is caused by the burrowing mite *Sarcoptes scabiei*, which has caused epizootics involving high morbidity and mortality rates in some wild mammal populations, mainly ungulate species [1,2]. Different host species, and even individuals within the same species, show variations in the severity and location of the lesions. For example, studies in the Southern UK [3] and Northern Spain [4,5] have reported that the disease manifests predominantly as hyperkeratotic lesions (type I hypersensitivity) in wild ungulates such as chamois and red deer, but as alopecic lesions (type IV hypersensitivity) in the wolf and fox. Nevertheless, hyperkeratotic lesions are also commonly observed in foxes [3,5,6].

Species and individual differences have been attributed to differences in the immune response [7], clinical stage [6,8] and concomitant presence of immunosuppressive pathogens [9]. In fact, *S. scabiei* itself can modulate various aspects of the mammalian innate and adaptive immune responses [10]. Identifying what immune responses are triggered upon infestation with *S. scabiei* and their relative efficacy may improve our understanding of disease course. The most common immune response in ectoparasite-associated systemic or local dermatitis in response to mites and its products is recruitment of eosinophils together with mast cells, typical of type I hypersensitivity [11]. Langerhans cells in the epidermis internalize sarcoptic antigen and migrate to regional lymph nodes, where they stimulate T cells [12]. In humans, the combined action of macrophages, T lymphocytes, and eosinophils can limit mite numbers and lesions [13]. Other inflammatory cells observed in mange lesions are plasma cells and B lymphocytes, which produce immunoglobulins that participate in the humoral immune response [14,15].

The main goal of this study was to analyze relative proportions of macrophages, plasma cells, as well as T and B lymphocytes in the skin of the wolf, fox, chamois and red deer from Asturias (Northern Spain) that were naturally infested with *S. scabiei*. The results may help us understand why infestation leads to different skin lesion patterns in certain species, which in turn may help guide efforts to manage the disease, i.e., treatment following capture in the wild of most susceptible species.

2. Materials and Methods

2.1. Samples

In previous work skin samples were taken from two wild carnivore species, the wolf (*Canis lupus*) and fox (*Vulpes vulpes*), and from two wild ruminant species, chamois (*Rupicapra rupicapra*) and red deer (*Cervus elaphus*), all from Asturias (Northern Spain) [5]. Afterwards, skin samples were paraffin-embedded, stained with hematoxylin–eosin and found to have natural sarcoptic mange, which was confirmed by mite isolation and identification. Gross examination showed that the wolf lesions were alopecic, while the fox, chamois and red deer lesions were extensive and hyperkeratotic. *Sarcoptes scabiei* mite burden was high in all species except the wolf. Eosinophils were not observed in wolf skin samples but they were observed in the other species [5]. Mast cells were not observed in any species.

Skin samples from 20 animals (5 per species) were selected for immunohistochemistry in this work. Ethical permission was not required.

2.2. Immunohistochemistry

Serial 3-μm paraffin sections were used for immunohistochemical detection of four different antigens (Table 1) using the Avidin-Biotin Complex (ABC) method (Vector Laboratories, CA, USA). Briefly, the sections were deparaffinized, rehydrated and rinsed with tap water. Endogenous peroxidase was quenched by incubating sections in methanol containing 3% H_2O_2 for 10 min at room temperature, then they were washed with water for 10 min. Antigens were retrieved using epitope demasking (Table 1), and nonspecific binding was inhibited by incubating the sections for 20 min at room temperature with 10% normal horse serum (detection of CD3) or 10% normal goat serum (detection of Iba1, lambda chain or CD20) in Tris-buffered saline (TBS) containing 5 mM Tris•HCl (pH 7.6), 136 mM NaCl and 1% bovine serum albumin. Tissue sections were incubated overnight at 4 °C with commercial mono- or polyclonal primary antibodies (Table 1), and then washed three times with TBS. Samples were incubated for 30 min at room temperature with horse anti-mouse serum or goat anti-rabbit serum (1:200, Vector Laboratories; Table 1), washed three times with TBS and then incubated with the ABC kit in TBS for 30 min at room temperature.

Table 1. Immunohistochemical protocols used to characterize types of inflammatory cell in skin lesion biopsies.

Primary Antibody (Dilution)	Target Cell Type	Epitope Demasking	Biotinylated Secondary Antibody (Dilution)
Iba1 (WAKO 019_19741), rabbit polyclonal (1:1000)	Macrophage	Microwave in citrate (pH 6), 20 min	Anti-rabbit (1:200)
Lambda (Dako A0193), rabbit polyclonal (1:1000)	Plasma cell	Microwave in citrate (pH 6), 20 min	Anti-rabbit (1:200)
CD3 (Novocastra-CL-L-CD3-565), mouse monoclonal (1:500)	pan-T cell	Microwave in citrate (pH 6), 20 min	Anti-mouse (1:200)
CD20 (ThermoFisher-PA516701), rabbit polyclonal (1:200)	pan-B cell	Steamer in citrate (pH 6), 20 min	Anti-rabbit (1:200)

Finally, the sections were incubated for 5 min with the substrate 3,3'-diaminobenzidine tetrahydrochloride (DAB; Sigma, St. Louis, MO, USA) and washed with TBS and water. After staining for 45 s with hematoxylin, slides were dehydrated and mounted with DPX (Fluka, Sigma, St. Louis, MO, USA). Stained slides were studied under light microscopy (Olympus BH—2) and photographed using a digital camera (Olympus DP—12). Each immunohistochemical staining included a positive control, in which the target antigen was present in the control section and the specific antibody was used (Figure S1); as well as a negative control, in which the primary antibody was omitted.

2.3. Cell Counting and Statistical Analysis

A total of 80 slides (4 slides per animal, 5 animals per species) were used for cell counting. Cells positive for each immunostained marker were quantified in five fields of each slide at 400× magnification using an image analysis program (Imaging Software NIS-Elements 3.20, Nikon, Tokyo, Japan). Then the mean proportion of stained cells to total cells was averaged across the five fields.

Descriptive and inferential statistics were used to analyze the distribution of four types of inflammatory cells in skin mange lesions in the four species. Data were tested for normality using the Shapiro–Wilk test. Species differences in the total number of inflammatory cells in scabies skin lesions were assessed for significance using the non-parametric Kruskal–Wallis and pairwise comparison tests. The percentage of total cells that stained for each of the four inflammatory cell biomarkers (Iba1, lambda chain, CD3 and CD20) was compared within and between each species using a one-way ANOVA. When significant differences were found, the Tukey test for multiple comparisons was applied.

As appropriate, data were expressed as the mean and standard deviation, or as the median and interquartile range. Data were analyzed using SPSS 24 for Windows (IBM, Chicago, IL, USA). A significance level of 0.05 was applied.

3. Results

3.1. Total Number of Inflammatory Cells

Samples from all species showed high numbers of inflammatory cells, with the fox showing the highest number. Inflammatory infiltrate was significantly higher in the fox than in the wolf and chamois (Table 2). On the other hand, while intra-species variation in the number of inflammatory cells was low for the fox and red deer, two of the five wolf samples showed lower numbers, and one chamois sample showed a much higher number.

Table 2. Total numbers of inflammatory cells in skin mange lesions from four species.

Species	n	Mean ± SD *	Median	IQR
Wolf	5	1175.2 ± 135.9 [a]	1179	1044.0–1304.5
Fox	5	1636.4 ± 195.8 [ab]	1725	1431.5–1797.0
Chamois	5	1242.4 ± 232.4 [b]	1192	1088.5–1421.5
Red deer	5	1293.6 ± 137.3	1228	1206.5–1413.5

* SD, standard deviation. IQR, interquartile range. [a,b] Statistical analysis by pairwise comparisons showed significant differences between means followed by same letters in the same column ($p < 0.05$).

3.2. Relative Proportions of Inflammatory Cell Types within Each Species

In all species, the most abundant cells in the inflammatory infiltrate were macrophages, while the least abundant were T or B lymphocytes (Figure 1).

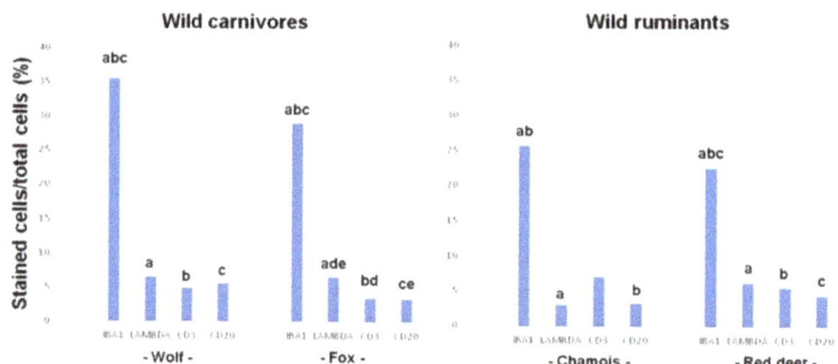

Figure 1. Intra-species differences in the numbers of macrophages (based on Iba1 immunostaining), plasma cells (lambda chain), T lymphocytes (CD3) and B lymphocytes (CD20) in skin mange lesions from wild carnivores and ruminants. For each animal species, the same letters above different rectangular bars indicates significant differences between means. For wolf: (a, b, c) $p < 0.001$. For fox: (a, b, c, d, e) $p < 0.05$. For chamois: (a, b) $p < 0.05$. For red deer: (a, b, c) $p < 0.001$. A Tukey test for multiple comparisons was applied for statistical analysis.

In the wolf and red deer, macrophages were significantly more abundant than plasma cells, T lymphocytes and B lymphocytes, which were similarly abundant (Figures 1 and 2). In the fox, the proportion of macrophages was significantly higher than the proportions of other cell types, and the proportion of plasma cells was significantly higher than those of T and B lymphocytes (Figure 1). In chamois, macrophages and T lymphocytes were more abundant than B lymphocytes and plasma cells (Figures 1 and 2), although the differences were significant only in the case of macrophages.

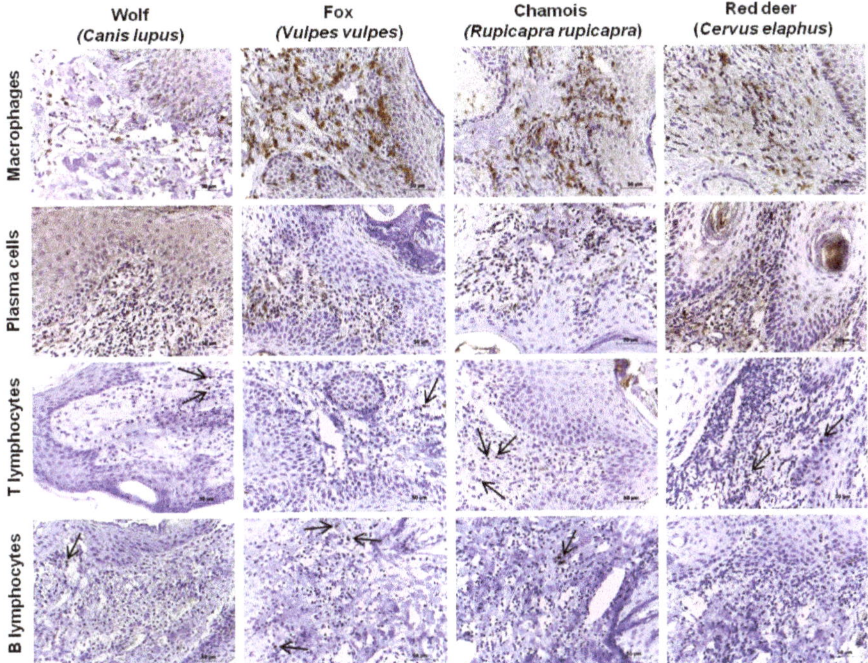

Figure 2. Comparative immunohistochemistry of cellular response in the dermis of wolves, foxes, chamois and red deer naturally infested with *Sarcoptes scabiei*. Skin biopsies were stained with primary antibodies against the indicated markers and the avidin-biotin complex kit. Macrophages were the predominant type of inflammatory cell in all species (top row). Plasma cells were present in all species, but less abundant than macrophages (second row). T lymphocytes (third row) and B lymphocytes (bottom row) were scarce in all species. Arrows indicate examples of stained cells when they are scarce (even with no B lymphocytes in red deer). Tissue sections in each column came from the same animal, and the results shown are representative of all five animals of each species. Bars = 50 microns.

3.3. Relative Proportions of Inflammatory Cell Types across Species

Wolves showed more abundant macrophages than the other species, although the difference was significant only with red deer (Table 3). Plasma cells were similarly abundant in the wolf, fox and red deer; the abundance in the wolf and fox was significantly higher than in chamois. Abundance of T and B lymphocytes was similarly low across all four species. Of all species, chamois showed the highest abundance of T lymphocytes and wolf showed the highest abundance of B lymphocytes, though the differences across the four species were not significant.

Table 3. Percentages of cells staining positive for inflammatory cell biomarkers in skin lesions from four species.

Biomarker (Target Cell Type)	Animal Species	% Positive Cells		
		n	Mean *	SD
Iba1 (macrophages)	Wolf	5	35.59 [a]	4.11
	Fox	5	29.11	7.94
	Chamois	5	25.70	9.60
	Red deer	5	22.57 [a]	3.96

Table 3. *Cont.*

Biomarker (Target Cell Type)	Animal Species	% Positive Cells		
		n	Mean *	SD
Lambda chain (plasma cells)	Wolf	5	6.61 [b]	0.97
	Fox	5	6.56 [c]	1.29
	Chamois	5	2.99 [bc]	0.28
	Red deer	5	6.31	3.19
CD3 (T lymphocytes)	Wolf	5	4.88	1.88
	Fox	5	3.39	1.53
	Chamois	5	7.15	3.59
	Red deer	5	5.59	1.49
CD20 (B lymphocytes)	Wolf	5	5.60	2.08
	Fox	5	3.33	0.90
	Chamois	5	3.31	0.93
	Red deer	5	4.42	1.87

* SD, standard deviation. Means followed by same letters in the same column differ significantly (Tukey test): [a] $p = 0.038$, [b] $p = 0.004$ and [c] $p = 0.017$.

4. Discussion

This study examined the immune response in the dermis of the wolf, fox, chamois and red deer from Northern Spain against natural *S. scabiei* infestation. We identified macrophages as the predominant cells in lesions of all four species, and we found small differences in the immune response among species.

Our observation that fox lesions contained the highest number of inflammatory cells may mean that the animals were in a more severe stage of the disease, or it may mean that their immune response of macrophages and lymphocytes was unable to control disease progression based on the generalized and crust lesions observed in those animals [6,8]. We cannot exclude that the higher cell number reflects higher parasitic burden [5], but this seems less likely, since chamois and red deer showed less intense inflammatory response despite high parasite burden [5]. Instead, the less intense response in the wolf, chamois and red deer may reflect some kind of adaptation or tolerance between the host and parasite [6,16]. These results should be verified and extended in further studies, especially since they varied appreciably among species and individuals within each species.

In all four species, macrophages were the most abundant inflammatory cells in skin lesions, yet the proportions varied across species. Lesions from the wolf showed a higher percentage of macrophages than lesions from other species, while lesions from fox and chamois showed a higher percentage of plasma cells and T lymphocytes, respectively. This may help explain species differences in hypersensitivity responses [5,16]. Ungulates and some fox populations tend to show immediate type I hypersensitivity response with higher eosinophil counts, while the wolf and other fox populations tend to show delayed type IV hypersensitivity response with higher macrophage counts [5]. Higher macrophage count may help explain the greater efficacy of a type IV hypersensitivity (alopecic) response in the wolf for eliminating *S. scabiei* [5]. The lower percentage of macrophages in foxes, chamois and red deer in the present study, together with the higher number of eosinophils previously observed in mange lesions [5], may result primarily in hyperkeratotic lesions related to a late phase of the type I hypersensitivity response pointing towards type IV hypersensitivity [6].

The higher proportion of plasma cells in fox than those of T and B lymphocytes might suggest a stronger humoral response [17]. In contrast, chamois in our study showed a larger proportion of T lymphocytes than B lymphocytes and plasma cells, consistent with previous work [8], confirming that in this species the cellular immune response is much stronger than the humoral response to *S. scabei* infestation [17]. That response may be related to the severe hyperkeratotic type lesions in our animals, which might reflect the ineffective T helper 2 (Th2) lymphocyte-type immune response [8]. In fact, excessive signaling by Th2 lymphocytes can trigger atopies or allergies [12]. The relative proportions of the different types of inflammatory cells may translate to differences in immune efficacy against *S. scabiei* infestation. Future studies should explore the factors that influence the nature and efficacy of the immune response against *S. scabiei*, which likely include body condition, sex, age, clinical stage, concomitant presence of immunosuppressive pathogens and sampling season [9]. Our results illustrate how immunohistochemistry of skin samples from wild species affected by sarcoptic mange can be useful for analyzing the immune response to infestation.

5. Conclusions

Our studies of four host species indicated a low proportion of B lymphocytes, T lymphocytes and plasma cells and a high proportion of macrophages in response to *S. scabiei* infestation, suggesting that these species mount a primarily innate immune response and are relatively poor at developing an adaptive immune response to this pathogen. Our findings further suggest that sarcoptic mange skin lesions may reflect a substantial humoral immune response in fox or a cellular immune response in chamois. Our observation of highest macrophage abundance in wolves may help explain their apparently more effective immune response against *S. scabiei*.

Supplementary Materials: The following are available online at http://www.mdpi.com/2076-2615/10/7/1146/s1, Figure S1: Immunohistochemical technique in positive controls; lymph node from badger.

Author Contributions: Conceptualization, A.B.; methodology, I.Z.M., Á.O., I.S., C.P.-M., M.J.G.-I., J.F.G.M., and A.B.; validation, C.P.-M., M.J.G.-I., J.F.G.M., and A.B.; formal analysis, I.Z.M., Á.O., C.P.-M., M.J.G.-I., J.F.G.M., and A.B; investigation, I.Z.M. and A.B.; resources, A.B; data curation, C.P.-M., M.J.G.-I., J.F.G.M., and A.B.; writing—original draft preparation, I.Z.M. and A.B.; writing—review and editing, I.Z.M., Á.O., C.P.-M., M.J.G.-I., J.F.G.M., and A.B. All authors have read and agreed to the published version of the manuscript.

Funding: This work was partially supported by the Principado de Asturias, PCTI 2018–220 (GRUPIN: IDI2018-000237 and FEDER). Ms. Ileana Z. Martínez was supported by a Fundación Carolina PhD scholarship (2017 call).

Acknowledgments: The authors would like to thank the Vice-Ministry of the Environment of the Principado of Asturias. The authors thank A. Chapin Rodríguez for critically reviewing the manuscript.

Conflicts of Interest: The authors declare no conflict of interest.

References

1. Fernández-Morán, J.; Gómez, S.; Ballesteros, F.; Quirós, P.; Benito, J.; Feliu, C.; Nieto, J. Epizootiology of sarcoptic mange in a population of cantabrian chamois (Rupicapra pyrenaica parava) in Northwestern Spain. *Vet. Parasitol.* **1997**, *73*, 163–171. [CrossRef]
2. León-Vizcaíno, L.; Cubero, M.; González-Capitel, E.; Simón, M.A.; Pérez, L.; Ruiz de Ybáñez, M.; Ortíz, J.M.; González, M.; Alonso, F. Experimental ivermectin treatment of sarcoptic mange and establishment of a mange-free population of Spanish ibex. *J. Wild. Dis.* **2001**, *37*, 775–785. [CrossRef] [PubMed]
3. Bates, P. Sarcoptic mange (Sarcoptes scabiei var vulpes) in a red fox (Vulpes vulpes) population in north-west Surrey. *Vet. Rec.* **2003**, *152*, 112–114. [CrossRef] [PubMed]
4. Oleaga, Á.; Casais, R.; Balseiro, A.; Espí, A.; Llaneza, L.; Hartasánchez, A.; Gortázar, C. New techniques for an old disease: Sarcoptic mange in the Iberian wolf. *Vet. Parasitol.* **2011**, *181*, 255–266. [CrossRef]
5. Oleaga, A.; Casais, R.; Prieto, J.M.; Gortázar, C.; Balseiro, A. Comparative pathological and immunohistochemical features of sarcoptic mange in five sympatric wildlife species in Northern Spain. *Eur. J. Wild. Res.* **2012**, *58*, 997–1000. [CrossRef]
6. Nimmervoll, H.; Hoby, S.; Robert, N.; Lommano, E.; Welle, M.; Ryser-Degiorgis, M.P. Pathology of sarcoptic mange in red foxes (Vulpes vulpes): macroscopic and histologic characterization of three disease stages. *J. Wild. Dis.* **2013**, *49*, 91–102. [CrossRef]
7. Oleaga, A.; García, A.; Balseiro, A.; Casais, R.; Mata, E.; Crespo, E. First description of sarcoptic mange in the endangered Iberian lynx (Lynx pardinus): Clinical and epidemiological features. *Eur. J. Wild. Res.* **2019**. [CrossRef] [PubMed]

8.   Salvadori, C.; Rocchigiani, G.; Lazzarotti, C.; Formenti, N.; Trogu, T.; Lanfranchi, P.; Zanardello, C.; Citterio, C.; Poli, A. Histological lesions and cellular response in the skin of Alpine chamois (Rupicapra r. rupicapra) spontaneously affected by sarcoptic mange. *Biomed. Res. Int.* **2016**, *2016*, 1–8. [CrossRef] [PubMed]

9.   Oleaga, A.; Vicente, J.; Ferroglio, E.; Pegoraro de Macedo, M.R.; Casais, R.; del Cerro, A.; Espí, A.; García, E.J.; Gortázar, C. Concomitance and interactions of pathogens in the Iberian wolf (Canis lupus). *Res. Vet. Sci.* **2015**, *101*, 22–27. [CrossRef] [PubMed]

10. Arlian, L.G.; Morgan, M.S. A review of Sarcoptes scabiei: Past, present and future. *Parasite Vector.* **2017**, *10*, 297. [CrossRef] [PubMed]

11. Bornstein, S.; Mörner, T.; Samuel, W.M. Sarcoptes scabiei and sarcoptic mange. In *Parasitic Diseases of Wild Mammals*; Samuel, W.M., Pybus, M.J., Kocan, A.A., Eds.; Iowa State University Press: Ames, IA, USA, 2001; pp. 107–109.

12. Lalli, P.N.; Morgan, M.S.; Arlian, L.G. Skewed Th1/Th2 immune response to Sarcoptes scabiei. *J. Parasitol.* **2004**, *90*, 711–714. [CrossRef] [PubMed]

13. Walton, S.F. The immunology of susceptibility and resistance to scabies. *Parasite Immunol.* **2010**, *32*, 532–540. [CrossRef] [PubMed]

14. Rambozzi, L.; Menzano, A.; Lavín, S.; Rossi, L. Biotin-avidin amplified ELISA for detection of antibodies to Sarcoptes scabiei in chamois (Rupicapra spp.). *Vet. Res.* **2004**, *35*, 701–708. [CrossRef] [PubMed]

15. Sarasa, M.; Rambozzi, L.; Rossi, L.; Meneguz, P.G.; Serrano, E.; Granados, J.E.; González, F.J.; Fandos, P.; Soriguer, R.C.; González, G.; et al. Sarcoptes scabiei: Specific immune response to sarcoptic mange in the Iberian ibex Capra pyrenaica depends on previous exposure and sex. *Exp. Parasitol.* **2010**, *124*, 265–271. [CrossRef] [PubMed]

16. Skerratt, L.F. Cellular response in the dermis of common wombats (Vombatus ursinus) infected with Sarcoptes scabiei var. wombati. *J. Wild. Res.* **2003**, *39*, 193–202. [CrossRef] [PubMed]

17. Arlian, L.G.; Morgan, M.S.; Rapp, C.M.; Vyszenski-Moher, D.L. The development of protective immunity in canine scabies. *Vet. Parasitol.* **1996**, *62*, 133–142. [CrossRef]

Article

Pathological Study of Facial Eczema (Pithomycotoxicosis) in Sheep

Miguel Fernández [1,2,*], Valentín Pérez [1,2], Miguel Fuertes [3], Julio Benavides [2], José Espinosa [1,2], Juan Menéndez [4,†], Ana L. García-Pérez [3] and M. Carmen Ferreras [1,2]

[1] Departamento de Sanidad Animal, Facultad de Veterinaria, Universidad de León, C/Prof. Pedro Cármenes s/n, E-24071 León, Spain; vperp@unileon.es (V.P.); jespic@unileon.es (J.E.); mcfere@unileon.es (M.C.F.)

[2] Instituto de Ganadería de Montaña (IGM), CSIC-Universidad de León, Finca Marzanas s/n, E-24346 León, Spain; julio.benavides@csic.es

[3] Department of Animal Health, NEIKER-Basque Institute for Agricultural Research and Development, Basque Research and Technology Alliance (BRTA), Parque Científico y Tecnológico de Bizkaia, P812, E-48160 Derio, Spain; mfuertes@neiker.eus (M.F.); agarcia@neiker.eus (A.L.G.-P.)

[4] Area de Sistemas de Producción Animal, Servicio Regional de Investigación y Desarrollo Agroalimentario, SERIDA, E-33300 Villaviciosa, Asturias, Spain; albeitar@centroveterinarioalbeitar.com

[*] Correspondence: m.fernandez@unileon.es; Tel.: +34-987-291232

[†] Current address: Centro Veterinario Albéitar, E-33204 Gijón, Spain.

Citation: Fernández, M.; Pérez, V.; Fuertes, M.; Benavides, J.; Espinosa, J.; Menéndez, J.; García-Pérez, A.L.; Ferreras, M.C. Pathological Study of Facial Eczema (Pithomycotoxicosis) in Sheep. *Animals* **2021**, *11*, 1070. https://doi.org/10.3390/ani11041070

Academic Editors: Alejandro Suárez-Bonnet and Gustavo A. Ramírez Rivero

Received: 19 March 2021
Accepted: 6 April 2021
Published: 9 April 2021

Publisher's Note: MDPI stays neutral with regard to jurisdictional claims in published maps and institutional affiliations.

Simple Summary: Facial eczema (FE) is a secondary photosensitization disease of farm ruminants caused by the sporidesmin A, present in the spores of the saprophytic fungus *Pithomyces chartarum*. This study communicates an outbreak of ovine FE in Asturias (Spain) and characterizes the local immune response that may contribute to liver damage promoting cholestasis and progression towards fibrosis and cirrhosis. Animals showed clinical signs of photosensitivity and lower gain of weight, loss of wool and crusting in the head for at least 6 months after the FE outbreak. Some sheep presented acute lesions characterized by subcutaneous edema in the head, cholestasis and nephrosis with macrophages and neutrophils present in areas of canalicular cholestasis. In chronic cases, alopecia and crusting, hepatic atrophy with regenerative nodules, fibrosis and gallstones were seen. The surviving parenchyma persisted with a jigsaw pattern characteristic of biliary cirrhosis. Concentric and eccentric myointimal proliferation was found in arteries near damaged bile ducts, where macrophages and lymphocytes were also observed.

Abstract: Facial eczema (FE) is a secondary photosensitization disease of farm ruminants caused by the sporidesmin A, produced in the spores of the saprophytic fungus *Pithomyces chartarum*. This study communicates an outbreak of ovine FE in Asturias (Spain) and characterizes the serum biochemical pattern and the immune response that may contribute to liver damage, favoring cholestasis and the progression to fibrosis and cirrhosis. Animals showed clinical signs of photosensitivity, with decrease of daily weight gain and loss of wool and crusting for at least 6 months after the FE outbreak. Serum activity of γ-glutamyltransferase and alkaline phosphatase were significantly increased in sheep with skin lesions. In the acute phase, edematous skin lesions in the head, hepatocytic and canalicular cholestasis in centrilobular regions, presence of neutrophils in small clumps surrounding deposits of bile pigment, ductular proliferation, as well as cholemic nephrosis, were observed. Macrophages, stained positively for MAC387, were found in areas of canalicular cholestasis. In the chronic phase, areas of alopecia and crusting were seen in the head, and the liver was atrophic with large regeneration nodules and gallstones. Fibrosis around dilated bile ducts, "typical" and "atypical" ductular reaction and an inflammatory infiltrate composed of lymphocytes and pigmented macrophages, with iron deposits and lipofuscin, were found. The surviving parenchyma persisted with a jigsaw pattern characteristic of biliary cirrhosis. Concentric and eccentric myointimal proliferation was found in arteries near damaged bile ducts. In cirrhotic livers, stellated cells, ductular reaction, ectatic bile ducts and presence of M2 macrophages and lymphocytes, were observed in areas of bile ductular reaction.

Keywords: facial eczema; sheep; pathology; pithomycotoxicosis; immunohistochemistry; cirrhosis; cholestasis; liver

1. Introduction

Facial eczema (pithomycotoxicosis) (FE) is a secondary-hepatogenous photosensitization disease of farm ruminants caused by the epipolythiodioxopiperazine mycotoxin sporidesmin A, present in the spores of the saprophytic fungus *Pithomyces chartarum* [1,2]. This fungus grows on dead vegetable matter at the base of the ryegrass-dominant pasture in all temperate worldwide zones during cloudy days with rain, with temperatures above 16 °C (optimal 24 °C), and relative humidity upper than 80% [3]. The portal circulation is the main route by which sporidesmin A enters to the liver following absorption from the intestinal tract [4]. In the biliary system, the sporidesmin A, characterized by the presence of an internal disulphide bridge, leads to the formation of toxic free-radicals that react with molecular oxygen to produce superoxide radicals [2,5,6]. The damaged liver is unable to remove a normal end product of chlorophyll metabolism, phylloerythrin (a photodynamic agent), from the blood for excretion in the bile [5]. When ruminants are exposed to sunlight, this pigment is responsible for unpigmented skin lesions [7]. FE was first recognized in New Zealand where it occurs more frequently [7]. However, this mycotoxic disease has also been reported in South Africa [8], Australia [9], the United States [10], France [11], Portugal [12], the Netherlands [13], Turkey [14], Uruguay and Argentina [3]. In Spain, the first and only reported outbreak occurred in the Basque Country [15]. In live animals, increases in the serum concentration of several enzymes such as gamma-glutamyltransferase (GGT) were found to be positively correlated to cholestasis and are indicators of bile duct damage in sporidesmin natural and experimental intoxication [1,7,12,16–19].

FE commonly affects sheep and cattle [7] while goats are more resistant to sporidesmin toxicosis than sheep [1]. The experimental sporidesmin toxicity in the rabbits has been demonstrated [17]. Although there was variation in susceptibility between individuals, the degree of liver injury and photosensitization appears to increase with both dose and length of time during which sporidesmin was administered [20]. The toxicity of pastures depends on the number of *P. chartarum* spores in the dead plant material and the toxicity of the particular *P. chartarum* strains [3]. Some authors reported that the most severe liver injury was due to the higher total consumption of spores caused by the combination of pasture spore concentration and dry matter intake [21]. Field observations reported that high spore counts (more than 40,000 spores/g of grass) of *P. chartarum* in grass samples can cause clinical signs [12], and there was a strong relationship between spore counts in ruminal content and severity of clinical signs [14]. In sheep, the clinical signs of photosensitization (erythema, edema and alopecia in unpigmented skin) appeared 14–18 days after intake of the toxin [1,20]. Jaundice [3] and loss of body weight in severe and chronic FE [7,20] are evident in this disease. In acute cases the liver is enlarged and shows a yellowish discoloration [10], and the gallbladder and extrahepatic bile ducts are distended [17]. In cases of longer evolution, there is liver atrophy and fibrosis [3] leading to liver cirrhosis [8,15].

The histological lesions include acute necrotizing cholangitis [12,17], bile stasis [1] and later on, bile duct hyperplasia and portal fibrosis [3,10,15]. Liver regeneration [7] and interlobular cirrhosis [8] have also been shown in the chronic phase of FE.

A lymphocytic infiltrate has been observed associated with the chronic hepatic changes of FE [8,10]. The main cell target in cholangiopathies are the epithelial cells lining the bile ducts (i.e., cholangiocytes), that are exposed to cytokines and inflammatory mediators produced by infiltrating lymphocytes, macrophages and activated myofibroblasts [22]. "Typical cholangiocyte proliferation" is usually found in acute obstructive cholestatic liver disease and "atypical ductular reaction" is commonly seen in human primary biliary cirrhosis [23]. Bile duct lesions have been described in subacute FE in cattle [24], but there are no studies about the type of ductular response in chronic FE cases or on how ductular cells interact with other cell types such as Kupffer and hepatic stellate cells (HSCs). Quiescent HSCs are resident perisinusoidal cells in the subendothelial space between hepatocytes and sinusoidal endothelial cells [25]. They are the primary site for storing retinoids (vitamin A) within the body. During liver injury HSCs proliferate and differentiate into contractile and matrix-producing myofibroblasts that generate progressive fibrosis and promoted a chemotactic activity for monocytes and lymphocytes, among others [25–27]. In liver pathology, ample evidence has been provided for an indirect role of macrophages in the development of fibrosis [25].

The main objective of this study is to characterize the inflammatory infiltrate, as well as the presence of HSCs and their distribution in damaged hepatic tissue in natural cases of FE from an outbreak that occurred in Northwestern Spain. Additionally, the acute and chronic liver lesions found in these cases of the disease, with special reference to the ductular reaction, are characterized, mainly considering their relationship with the presence of inflammatory cells, as a possible indicator of the role that the immune response can play in the development of liver damage in this toxicosis.

2. Materials and Methods

2.1. Ethical Information

Experimental animals were not used in this work. An observational study was performed with the blood samples obtained during regular veterinary clinical services and with *post mortem* tissue samples that are routinely collected after the death of animals.

2.2. Case History and Clinical Observations

The outbreak of FE occurred in La Mata, Grado, an inland municipality in the Principality of Asturias, in the north-west of Spain, between the end of September and the beginning of October 2003. This estate belonged to Servicio Regional de Investigación y Desarrollo Agroalimentario (SERIDA), Asturias. According to official data of the Asturian Society of Economic and Industrial Studies and the Meteorological Territorial Center of Asturias, the summer climate of 2003 was dry and very warm, with temperatures higher than 30 °C in the inland [28]. The autumn was warm (with temperatures above 19 °C in September and the first 19 days of October) and very rainy with a total monthly rainfall of 141.4 and 219 L/m^2 in October and November, respectively. The affected sheep were grazing on two plots (named 1B and "mixtures") at 50 m altitude. The plot 1B, of a 3.2 hectare (ha) of land, was sowed in 2001 with perennial ryegrass (*Lolium perenne* L. var. Tove), 30 kg/ha; hybrid ryegrass (*Lolium boucheanum* var. Kunth), 12 kg/ha and white clover (*Trifolium repens* var. Huia), 3 kg/ha. The plot mixtures of 2.5 ha was sowed in 1993 with perennial ryegrass (*Lolium perenne* L. var. Phoenix), hybrid ryegrass (*Lolium boucheanum* variety Dalita), and white clover (*Trifolium repens* var. Huia), at the same doses. In the plot 1B 28 crossbreed female sheep and 11 ewe lambs were grazing, of which 7 (6 adult and 1 ewe lamb) showed clinical signs of photosensitivity (pruritus, erythema and alopecia) on the face and ears. In the plot mixtures 2 adult sheep (over a total of 24) and 3 ewe lambs (over a total of 11) developed similar clinical signs, according to the information supplied by the practitioner. The four most affected animals had been treated symptomatically with Alergia-N (Pfizer), an antihistamine drug (cyprohetadine chlorhydrate, chlorphenamine maleate), and Penbex (Industrial veterinaria, S.A-INVESA), for treatment of secondary bacterial infections caused by germs sensitive to the association penicillin-dihydrostreptomycin sulfate, both by intramuscular route. Initially, the face skin lesions recovered when the animals were removed from the pastures for some days into the shade, but began to appear when the sheep returned to the plots and were exposed to sunlight. In the following 6 months, animals involved in the FE outbreak 4 sheep and 1 ewe lamb (plot 1B) and 1 sheep and 2 ewe lambs (plot mixtures) showed lower weight gains, wool loss and crusting in the dorsum of the head, nose and ears were observed.

2.3. Biochemical Assay and Statistical Analysis

Blood serum samples were taken from the jugular vein from 73 sheep (11 with and 62 without skin lesions) during the chronic stage of the episode (six months after the FE outbreak). Serum concentrations of γ-glutamyltransferase (GGT), alkaline phosphatase (ALP), aspartate aminotransferase (AST), albumin and total protein (TP) were determined on a multianalyser (Cobas Integra 400, Roche diagnostics). The results of the serum parameters analyzed were reported as mean, standard deviations and range (minimum and maximum), calculated using routine descriptive statistical procedures. The Kolmogorov–Smirnov test was used to assess normality of data. Non-parametric statistical methods were used to compare groups. Mann-Whitney U test was employed to compare the exposed animals that not showed cutaneous clinical signs with exposed sheep that presented skin lesions. p-values of less than 0.05 were considered statistically significant. All the statistical analyses were performed with the R software version 3.6.1 (R Development Core Team, R Foundation for Statistical Computing, Vienna, Austria, December 2019).

2.4. P. chartarum Spore Counts

A total of 36 grass samples from every field were randomly collected to count *P. chartarum* spores/g on 23 November 2003. Each grass sample was taken at least 10 m apart. The grass was cut 1 cm above the ground, avoiding taking soil, and cut into pieces of approximately 4 cm in length. From this mixture, 15 g of grass was taken and 150 mL of water was added, and then the mixture was homogenized for 3 min, in order to release the spores in the water. The *P. chartarum* spores present in each sample were identified and counted in an aliquot of wash water, using a Fusch-Rossental chamber. Finally, depending on the volume of water investigated, calculations were made to express the results as number of spores/g of grass. These grass samples were analysed in the Department of Animal Health of NEIKER- Basque Institute for Agricultural Research and Development.

2.5. Animal Cases and Pathological Examination

One adult sheep (plot 1B) with clinical signs suffering the acute phase of the disease, two adult animals (1 of the plot 1B and 1 of the plot mixtures), two months after the FE outbreak, and two adult sheep of the plot 1B six months after the FE outbreak, all of them belonging to the Galician breed, were examined in this study. Two sheep of a different estate were selected as reference and healthy animals. All of them were submitted to the Pathologic Diagnostic Service of the Veterinary Faculty of León over a seven-month period (November 2003–May 2004). In November an adult sheep with clinical signs in the acute phase of FE was humanely euthanized in the flock and were submitted for necropsy. Only when the pathological findings were discussed with the clinician, there was enough evidence to consider sporidesmin toxicosis as a possible cause of the liver and skin lesions. Two affected sheep and two other healthy control sheep were examined at slaughter in January and different tissue specimens (liver, skin, kidney, lung, heart) were submitted for histopathology. Finally, in May, two adult alive sheep were submitted for necropsy and euthanasia was performed by intravenously injection of barbiturate (T61; Intervet International, Madrid, Spain), after xylazine (Bayer, Leverkusen, Germany) subcutaneous administration, followed by exsanguination after the severing of carotid arteries and jugular veins according to our institution guidelines. Complete necropsies were performed in 3 sheep, one submitted during the acute, and two during chronic phases of FE. After gross examination, representative tissue samples (skin, liver, kidney, spleen, mesenteric lymph nodes, intestine, pancreas, adrenal glands, heart, lung, skeletal muscle and brain), were collected in all necropsied animals. All samples were fixed in 10% neutral-buffered formalin, processed routinely and embedded in paraffin wax. Sections (4 μm) were cut, mounted on glass microscope slides and stained with haematoxylin and eosin (HE), Masson Goldner trichrome for collagen, AFIP (Armed Forces Institute of Pathology) method for lipofuscin, Hall´s bilirubin stain, Perls´ Prussian blue stain for ferric pigments and acid rubeanic method for copper. Two livers with lesions consistent with steatosis and macronodular cirrhosis belonging to sheep of plot 1B that died in January and May, respectively, were not included in this study due to severe *post mortem* alterations (autolysis and putrefaction).

2.6. Immunohistochemistry

Selected sections (4 μm) from the liver were immunohistochemically labelled with a panel of antibodies. In all the cases, a polymer-based detection system (EnVision® System Labelled Polymer-HRP; Dako, Glostrup, Denmark) was employed, following the manufacturer instructions. Subsequently, immunolabelling was developed with a solution of 3.3′diaminobenzidine (DAB) or AP solution (Vector Laboratories, Burlingame, CA, USA). The slides were counterstained with Mayer's haematoxylin and mounted in hydrophobic medium (Table 1).

Table 1. Antibodies, specificity and immunohistochemical procedure used.

Antibody	Clone	Type	Marker For	Antigen Retrieval	Dilution	Source
CD3	-	Rabbit, Policlonal	T cells	PTLink/pH6/20′	1:300	Dako, Denmark
IgG	-	Biotinylated antibody	Plasma cells	PTLink/pH9/20′	1:200	Vector Lab, USA
Calprotectin	MAC387	Mouse, Monoclonal	Macrophages, activated epithelial cells	PTLink/pH9/20′	1:200	Gene Tex, USA
Lysozyme	-	Rabbit, Policlonal	Macrophages	PTLink/pH6/20′	1:250	Dako, Denmark
CD 206	MR5D3	Rat, Monoclonal	Macrophages (mannose receptor)	PTLink/pH6/20′	1:100	Gene Tex, USA
TGF β	TGFB-1	Mouse, Monoclonal	Hepatic stellate cells, Kupffer cells	PTLink/pH6/20′	1:200	GeneTex, USA
α-SMA	1A4	Mouse, Monoclonal	Smooth muscle cells, Myofibroblasts, HSCs	PTLink/pH9/20′	1:100	Dako, Denmark
Pankeratin	Pck26	Mouse, Monoclonal	Epithelial cells	Trypsin, 15′	1:200	Dako, Denmark

3. Results

3.1. Serum Biochemistry

Table 2 showed the mean ± standard deviation and range (minimum-maximum) of albumin, TP, AST, ALP and GGT concentrations according to the clinical status: exposed animals that not showed cutaneous clinical signs (a) and exposed sheep that presented skin lesions (b). Serum chemistry references values in sheep are shown in Table 2 for each biochemical parameter analyzed (c) [29].

Table 2. Serum liver enzymes in sheep with and without cutaneous lesions 6 months after the onset of the facial eczema (FE) outbreak. Different superscripts between columns shown significant differences between the animal groups compared ($* p < 0.05$; $** p < 0.01$; $*** p < 0.001$).

	Albumin (g/dL)	TP (g/dL)	AST (U/L)	ALP (U/L)	GGT (U/L)
(a) No Lesion ($n = 62$)	3.72 ± 0.54 [a] ** (1.69–3.86)	7.20 ± 0.80 [a] * (4.81–9.39)	237.82 ± 142.76 [a] (89.7–762.6)	283.39 ± 180.06 [a] ** (34–954.9)	156.15 ± 266.28 [a] *** (40.1–1838.30)
(b) Skin lesion ($n = 11$)	2.91 ± 0.51 [b] (2.20–6.17)	8.05 ± 1.09 [b] (6.80–10.34)	244.4 ± 105.60 [a] (98.10–408.20)	587.35 ± 447.37 [b] (127–1444.40)	628.16 ± 412.69 [b] (64.80–1348.20)
(c) Reference values range	2.70–3.70	5.90–7.80	49–123.30	26.90–156.10	19.60–49.10

3.2. P. chartarum Spore Counts

Mean ± standard deviation and range (minimum-maximum) of the number of *P. chartarum* spores per gram of pasture in the 36 grass samples analyzed, was 11,389 ± 15,100 (0–75,000) spores/g of grass.

3.3. Pathology

3.3.1. Acute Stage

Animals examined at this stage of the disease showed edematous drooping ears, serum exudation and sloughing of the skin in the dorsum of the ears, eyelids, muzzle, lips and forelimbs, characterized histologically by epidermal necrosis and presence of serocellular crusts (Figure 1A). The carcass showed intense jaundice and the liver were bile stained (yellowish) and enlarged (Figure 1B). The extrahepatic bile ducts and the gallbladder were distended by abundant bile and visible brown concretions (pigmented calculi) (Figure 1B). Histologically, in the liver, there was a macrovesicular (large droplet) predominantly periportal fatty change (zone 1). Hepatocytic and canalicular cholestasis with bile granular deposits in the cytoplasm of hepatocytes and presence of bile plugs in canaliculi respectively, confirmed by the Hall histochemical stain, were prominent in centrilobular regions (zone 3). Some of these cells were swollen or necrotic. Small clumps of polimorphonuclear neutrophils (PMNs), Kupffer cells (KCs) with AFIP lipofuscin positive granules, and foreign-body-type giant cells were present surrounding more prominent deposits of bile pigment ('bile lakes') (Figure 1C). Portal tracts were edematous with ductular proliferation at the periphery (bile ducts with a well-defined lumen). There was also a mild (sparse) portal inflammation (scattered lipofuscin-laden macrophages, detected by the AFIP method, and lymphocytes). The epithelial cells in septal bile ducts were shrunken with picnotic nuclei, and larger bile ducts were dilated and contained inspissated bile. The kidney was macroscopically slightly enlarged and showed a greenish discoloration with linear green streaks throughout the cortex and medulla (Figure 1D). Diffuse cellular swelling and green granular pigmented bilirubin, in the proximal tubules, as well as green-yellow acellular tubular bile casts in the distal nephron segments of the renal medulla, both confirmed by the Hall histochemical stain, were observed microscopically (cholemic nephrosis) (Figure 1E,F).

3.3.2. Chronic Stage

All the sheep submitted 2 and 6 months after the initial onset of the disease outbreak showed skin lesions limited to areas of the head: dorsum of the head and ears, eyelids, face, lips and nose were alopecic and crusting. In the ears, there was focal necrosis of the epidermis and the underlying cartilage with palisading crust formation. The livers were atrophic, with lesions most marked in the left lobe (2 sheep), yellowish with whitish bands of fibrous tissue and larger lonely nodules of 4 and 8 cm located principally in the visceral surface in the quadrate lobe (2 sheep) (Figure 2A). The intrahepatic and extrahepatic bile ducts and gallbladder were also enlarged and contained biliary sludge and pigmented gallstones.

Histologically, increased biliary fibrosis with an associated inflammatory infiltrate, and prominent ductular proliferation were observed in the liver of all sheep examined. The portal tracts were expanded with proliferating bile ductules with a well-defined lumen (typical cholangiocyte proliferation) and fibrous tissue, which occasionally bridged adjacent portal tracts. Extensive pericellular or subsinusoidal fibrosis was evident using Masson-Goldner trichrome stain. An intense leucocytic portal inflammation (lymphocytes and plasma cells) was noted. At the same time and in the same liver, areas formed by an irregular proliferation of intrahepatic bile ductules at the portal tract margins, with poorly formed lumina that replaces the hepatic parenchyma (atypical ductular reaction) (Figure 2B). Numerous spindle-shaped cells, lymphocytes and macrophages containing brown pigment, were seen throughout the fibrous tissue, as well as in remnants of the hepatic lobules which eventually disappeared, principally in the hepatic left lobe (Figure 2C).

Histochemical studies showed that pigmented macrophages in close association with remnants of hepatocytes were positively red-stained with AFIP method for lipofuscin. This pigmented lipoproteins coexisted in some macrophages with iron deposits (hemosiderin) that reacted with Perls´ Prussian blue stain (Figure 2D). Granular deposits of protein-bound copper salts were observed in periportal hepatocytes and pigmented macrophages as small black granules in their cytoplasm. Collagen fibers were arranged in concentric layers around dilated interlobular bile ducts and proliferating bile ductules ("onion skin" fibrosis). As the result of coalescence of adjacent fibrotic portal tracts, portal-portal fibrous septa were noticed and the surviving parenchyma persisted with a jigsaw pattern characteristic of the biliary cirrhosis (Figure 2E). Larger bile ducts in sheep examined two months after FE outbreak were dilated and contained inspissated bile plugs and biliary stones. The biliary epithelium was flattened and necrotic and the fibrotic wall contained numerous pigmented macrophages and foreign body type giant cells around bile pigment deposits, admixed with mononuclear inflammatory cells. A marked fibrotic thickening, with a mild lymphocyte infiltrate, was the major histologic finding in large intrahepatic and extrahepatic bile ducts 6 months after the FE outbreak. Canalicular cholestatic changes were only seen in one sheep 2 months after the outbreak. Additionally, this sheep with cholestasis also showed brownish discoloration of the renal cortex and medulla. Intracellular bile pigment, stained green by the Hall histochemical stain, involved proximal tubules in two sheep 2 months after the FE outbreak (cholemic nephrosis).

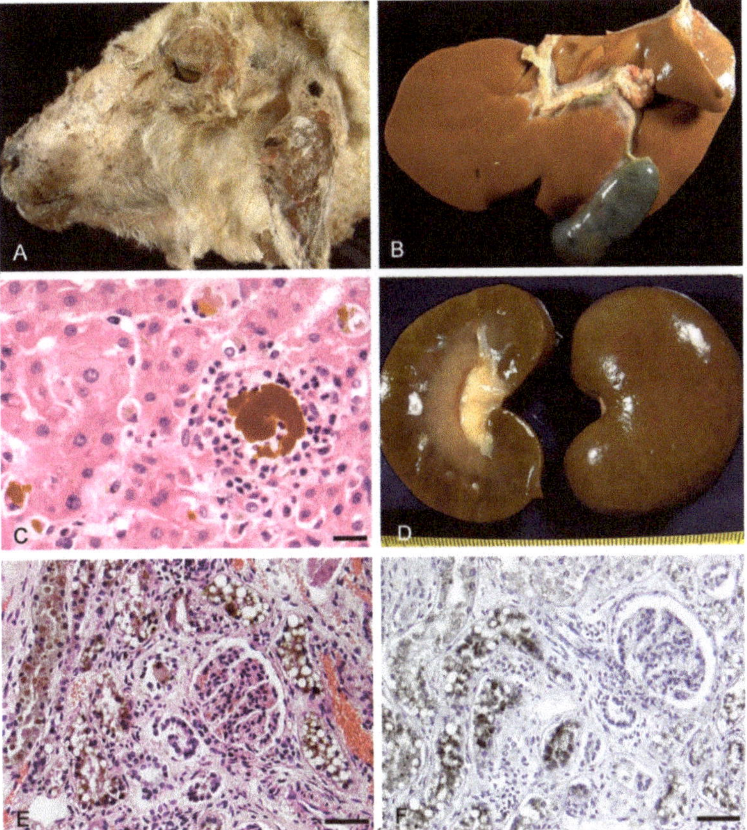

Figure 1. Acute lesions observed in naturally acquired cases of FE in sheep. (**A**) Sloughing of the skin in the ears, eyelids, muzzle and lips. (**B**) The liver was diffusely yellowish in color and the common and cystic bile duct were distended. (**C**) Polimorphonuclear neutrophils were surrounding a hepatic "bile lake". HE. Bar, 20 μm. (**D**) Linear greenish discoloration of the renal cortex and medulla. (**E**) Brown granular pigment in the proximal tubules. HE. Bar, 50 μm. (**F**) Green granular pigment consistent with bilirubin in the proximal tubules. Hall´s bilirubin stain. Bar, 50 μm.

All sheep examined in the chronic stage showed vascular lesions in the liver. In portal hepatic arteries, hepatic arteries in the septa and large hilar hepatic arteries, near severely damaged intrahepatic (interlobular, septal) and extrahepatic (common and cystic) bile ducts, both concentric and eccentric myointimal proliferation was found to some degree (Figure 2F). Similar occlusive lesions, characterized by intimal cap proliferation, were observed in sublobular veins near damaged bile ducts. Lymphocytic phlebitis and phlebosclerosis consisting, respectively, of chronic inflammatory infiltrate of the wall and perivenular fibrous thickening of central veins and striking fibromuscular hypertrophy of the walls of ectatic hepatic veins were also found.

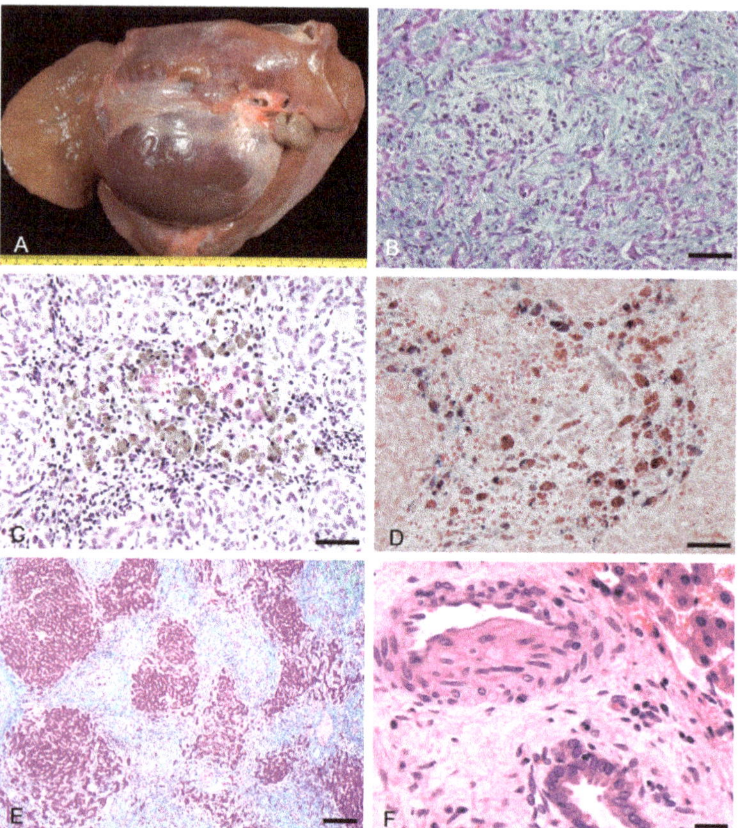

Figure 2. Chronic lesions observed in naturally acquired cases of FE in sheep. (**A**) Marked atrophy in the hepatic left lobe and a large nodule in the hepatic visceral surface. (**B**) Hepatic lobular areas are replaced by proliferated bile ductules (atypical ductular reaction) and fibrous tissue. Masson-Goldner trichome stain. Bar, 50 μm. (**C**) Numerous lymphocytes and pigmented macrophages were seen in the fibrous tissue in association with remnants of hepatic lobes. Masson-Goldner trichome stain. Bar, 50 μm. (**D**) Macrophages were positively red stained for lipofuscin (AFIP stain) that coexisted in some cells with hemosiderin. Perls' Prussian blue stain. Bar, 50 μm. (**E**) Jigsaw pattern characteristic of the biliary cirrhosis. Masson-Goldner trichome stain. Bar, 200 μm. (**F**) Eccentric myointimal proliferation in a hepatic arteriole adjacent to the bile duct. HE. Bar, 20 μm.

3.4. Immunohistochemistry

The epithelial cell marker (pan-cytokeratin) was strongly expressed in the epithelium of bile ducts and, with less intensity, in acute and chronic biliary ductular reaction. This antibody was especially useful for the identification of the bile duct epithelium of damaged bile ducts that developed granulomas composed of macrophages, as well as the ectatic bile ducts surrounded by fibrous tissue in portal tracts (Figure 3A,B).

In the acute phase, there was an increase of α-SMA-positive into the parenchyma in areas of canalicular cholestasis and hepatocyte necrosis as well as within portal tracts around proliferating typical bile ductules (Figure 3C). In chronic FE these cells showed strong immunopositivity and accumulate surrounding bile ductular structures and ectatic bile ducts in fibrotic portal tracts and septa (Figure 3D). In this last case α-SMA-positively immunolabelled HSCs, appeared as spindle-shaped cells that delimited both bile ducts and atypical ductules. α-SMA was not expressed in ductular cells, cholangyocytes or hepatocytes. An increase in the number of macrophages and PMNs that stained positively for both anti-lysozyme and anti-MAC387 antibodies was observed in the areas of canalicular cholestasis and hepatocyte necrosis in acute FE liver lesions (Figure 4A). Many cells with the characteristic morphology of KCs have strong cytoplasmic labelling for lysozyme, and MAC387 immunostaining clearly defined cell aggregates, with extensive immunoreactivity, scattered in the liver parenchyma, both in zones 2 and 3 of the hepatic lobes. In addition MAC387 positivity was observed in cells having the morphology of blood monocytes within blood vessels.

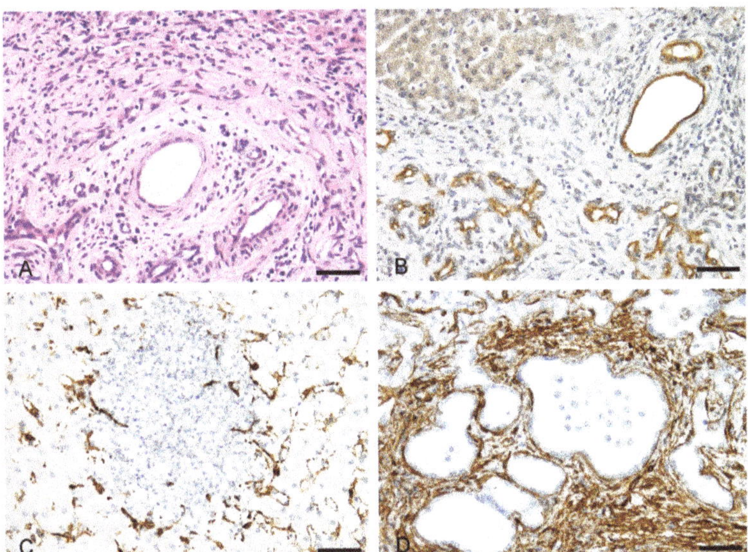

Figure 3. Immunohistochemistry in naturally acute and chronic acquired cases of FE in sheep. (**A**) Ectatic interlobular bile duct surrounded by a thin layer of collagen fibers ('onion skin fibrosis'). HE. Bar, 50 μm. (**B**) Pancytokeratin antibody stain positively ectatic bile ducts with flattened epthelium, similar to showed in Figure 3A, and bile ductules. Bar, 50 μm. (**C**) α-SMA-+ HSCs in areas of canalicular cholestasis in acute FE liver lesions. Bar, 50 μm. (**D**) In chronic liver lesions observed in FE, α-SMA-+ HSCs cells accumulated and surrounded ectatic bile ducts. Bar, 50 μm.

In the chronic hepatic lesions, abundant pigmented macrophages in close association with remnants of hepatocyte lobules, were positive for lysozyme and MAC387. Also, strong lysozyme and MAC387 positively immunolabelled macrophages were found as part of the granulomatous lesions formed around degenerated bile ducts. Remarkably, a decrease in the number of cells immunostained with these antibodies were observed in the fibrous septa in relation to atypical ductular proliferation. The mannose receptor (CD206) staining was only present in some pigmented macrophages observed as scattered cells in the portal tracts (acute phase), but were more numerous in chronic lesions, in fibrotic septa with marked ductular reaction (Figure 4B). No co-localization was found with lysozyme and MAC387 markers within cells in the hepatic lobes. It is remarkable that cells positively immunostained for CD206 antibody corresponded to macrophages showing intense TGF-β immunoreactivity. (Figure 4C). Sparse to intense intracellular staining for TGF- β was also observed in α-SMA-positive cells present in the thickened subendothelial areas and in the tunica media of hepatic arteries (Figure 4D).

In acute FE lesions, intrahepatic T CD3+-lymphocytes were scattered in sinusoids and portal tracts and occasionally were seen in intraepithelial location in bile ducts. In chronic FE lesions, there was a prominent T lymphocytic infiltrate forming aggregates intimately associated with the bile ducts, with a diffuse pattern or forming aggregates in areas of bile ductular reaction in the fibrous septa (Figure 4E). T CD3-positively immunostained lymphocytes were identified in granulomas around degenerate bile ducts and extravasated bile pigment as well as surrounding large bile ducts contain bile stones. IgG+ plasma cells were occasionally seen in portal tracts in acute FE liver lesions. In chronic cases, plasma cells were found scattered or in small amounts followed a similar distributional pattern than T lymphocytes, although they were less abundant. Like T lympohocytes, plasma cells were found in the concentric arrangement of fibrous tissue around bile ducts and ductules (Figure 4F).

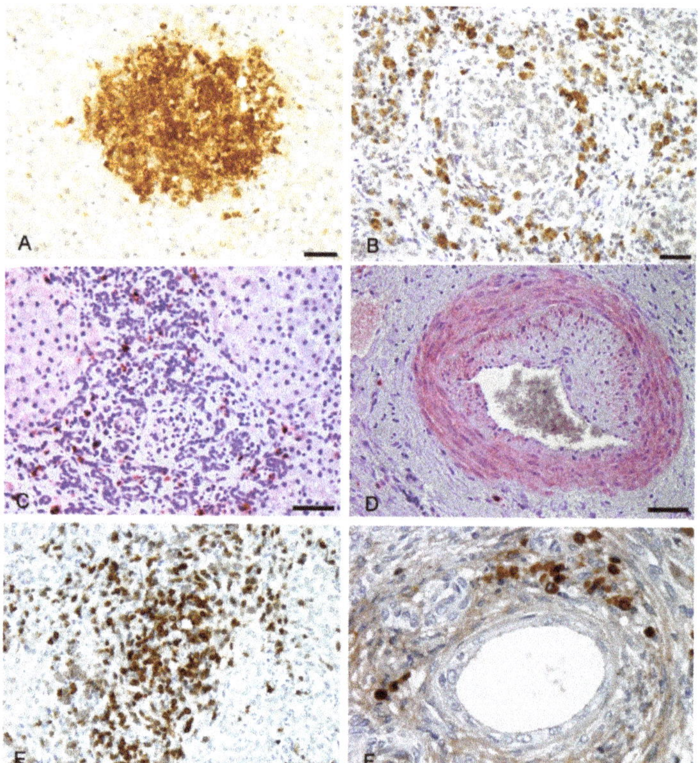

Figure 4. Immunohistochemistry in naturally acute and chronic acquired cases of FE in sheep. (**A**) Intense positive MAC387 immunolabeling of a cluster of macrophages and neutrophils in an area of cholestasis (acute lesion). Bar, 50 μm. (**B**) Presence of CD206-+ macrophages in areas of fibrosis and ductular reaction (DR) (chronic lesion). Bar, 50 μm. (**C**) Anti TGF-β antibody red stained cells morphologically compatible with macrophages in areas of DR (chronic lesion). Bar, 50 μm. (**D**) Positive immunolabeling for anti TGF-β antibody in a hepatic artery with an occlusive lesion (subintimal proliferation) in chronic FE cases. Bar, 50 μm. (**E**) Aggregates of T CD3-+ lymphocytes in areas of DR in chronic FE lesions. Bar, 50 μm. (**F**) In chronic FE lesions IgG-+ plasma cells were seen scattered around ectatic bile ducts. Bar, 20 μm.

4. Discussion

In this report we described an episode of FE in sheep that occurred in Asturias, Spain in 2003. This mycotoxic disease was first observed in European sheep in France [11] and later in the Basque Country, Spain [15] and the Azores, Portugal [12]. In late 2003, the weather conditions in Asturias were the most favorable for *P. chartarum* growth and sporulation. The autumn was warm (with temperatures above 19 °C in September and the first 19 days of October) and very rainy, with total monthly rainfall of 141.4 and 219 L/m^2 in October and November, respectively. Late summer and early autumn with warm temperatures (minimum 16 °C) and high humidity (above 90 per cent) were the periods favorable to FE outbreaks in the Azores, Portugal between 1999 to 2001 [12]. The periods and weather conditions in Asturias were similar to those observed in these islands, both with oceanic climate. Spore counts of *P. chartarum* identified in several different grass samples ranged from 0 to 75,000 spores/g of pasture. It has been suggested that spore counts as low as 50,000 spores/g of grass could be dangerous to livestock if grazed for long periods in sunlight and that the greater liver injury occurred when sheep were grazed on pastures having maximum spore counts of 130,000 spores/g of grass [20,21]. Other causes of toxins from fungi or plants were not found around the studied field. Given that FE clinical signs appear 14–18 days after intake of the mycotoxin [20] and, in the present outbreak, sheep with acute signs of photosensitization were observed from the middle of October, plots possibly reached the higher spore numbers in September/October. The low spore counts observed in the present study may be related to the variability between individual sites in the plot. Besides, the grass samples were collected in late November with a low temperature (average of 11.6 °C), conditions less favorable to fungal growth and sporulation.

In the present study, a significant rise in the activity of serum GGT associated with histological cholestasis and bile duct damage was demonstrated in sheep in accordance with previous reports in which sporidesmin was administered experimentally [16,30], and also in spontaneous intoxications [10,12,14,15]. GGT and ALP levels were higher in all sheep examined (when compared with reference ranges) and significantly increased ($p < 0.001$ and $p < 0.01$, respectively) in the 11 sheep with skin lesions when comparing with the 62 apparently healthy animals (without skin lesions). In our opinion, in agreement with previous reports [7,18], the detection of elevated GGT serum levels was the most suitable marker for monitoring FE affected sheep under field conditions because it correlated with the liver lesions, even in apparently healthy animals. In this sense, our results confirmed that a high activity of GGT persists for six months after an FE outbreak, in accordance with other published data [3]. Serum albumin was lower ($p < 0.01$) and TP was increased ($p < 0.05$) in sheep with skin lesions and the serum AST levels did not show statistically significant changes. AST was increased in all the tested animals, even in those without injury 6 months after the outbreak. This fact could be the consequence of a previous exposure and the development of a chronic inflammatory process, so that AST could be a sensitive and long-lived marker of liver damage in sheep, in disagreement with previous observations of natural intoxication with sporidesmin in sheep [18].

The gross lesions in the skin, liver and kidney described in the acute phase of ovine FE agree, in general, with previous descriptions in natural [10,14] and experimental cases [8,30]. Microscopic lesions described in the acute stage have been demonstrated experimentally in sheep dosed pure cultures of *P. chartarum* directly to the stomach, equivalent to approximately 3–4 mg/kg live mass [8,30]. These animals, that became photosensitive on the 9–10th day and died or were euthanized on the 4–10 subsequent days, showed hepatic parenchymal infarcts with leakage of bile between the necrotic hepatocytes and the presence of polymorphonuclear cells, similar to the microscopic findings described in the present study. A moderate bile duct proliferation and mononuclear cell infiltration (including some pigment laden macrophages), as well as necrotic lesions in the bile ducts in the portal tracts, were also observed in this work. Recently, altered cell adhesion and disruption of actin in sheep gallbladder epithelial cells incubated with sporidesmin were demonstrated, suggesting that the biliary tract pathology in FE may be due to the effects of the toxin on cytoplasmic and cell surface protein networks, affecting the integrity of the epithelial lining of the biliary tract [31]. The special stains used, for the first time in the present study, such as Hall and AFIP stains, were useful to demonstrate the presence of bile and lipofuscin, respectively, in liver. Lipofuscin can accumulate in liver macrophages in natural FE in sheep and may represent the remnants of phagocytosed debris from necrotic hepatocytes. This study also confirmed by the Hall stain the presence of cholemic nephrosis characterized by proximal tubulopathy and intrarenal bile cast formation in ovine FE. Previous studies showed that elevated plasma levels of conjugated bilirubin are related to renal failure associated with obstructive jaundice [32]. The term "bile cast nephropathy", caused by direct bilirubin toxicity and tubular obstruction, has been proposed for this pathologic entity in humans [33].

Liver atrophy, more severe in the left lobe, and large nodules of regeneration in the visceral surface of the liver, as well as alopecic and crusty head skin were the most striking gross findings in chronic forms of FE that are in agreement with previous descriptions [3,15,34].

The atrophic left hepatic lobe with dilated intrahepatic and extrahepatic bile ducts are conditions that may have been due to compression of the left trunk of the portal vein secondary to hepatolithiasis [35]. In our case, the mechanism of left lobar atrophy might have been due to the occlusion of the ducts by biliary sludge and pigmented gallstones that cause atrophy of the parenchyma served by them [34].

The presence of nodular regeneration and cirrhosis have been considered the most conspicuous pathological changes in natural cases of FE in sheep [8] and may start as early as 2 weeks after toxin insults [7]. Experimentally, this lesion has been observed in sheep challenged with a total of 2.125 mg/kg sporidesmin divided into 17 doses for 144 days [30].

A remarkable histological finding in chronic FE was a strong ductular reaction that replaces the hepatic parenchyma associated with extensive fibrosis and aggregates of lymphocytes and pigment containing macrophages in connective tissue septa. It has been suggested that this ductular reaction can represent regenerative proliferation of bipotential hepatic stem/progenitor cells that have the ability to differentiate into both hepatocytes and cholangiocytes, but there is no definitive evidence for it [23]. Pigmented macrophages contained deposits of lipofuscin, hemosiderin and copper salts, possibly as a result of an increase of their oxidative stress with iron-catalyzed production of reactive oxygen species causing oxidative damage to lipids and proteins [36]. In vitro, the autoxidation of reduced sporidesmin is catalyzed by iron and by copper and generate a dithiol, a superoxide free radical suggesting that any superoxide production from sporidesmin in vivo would be mediated by the intracellular transport pool of copper [37]. Lipofuscin and copper are also deposited in biliary cirrhosis and chronic cholestatic diseases, respectively [38].

In this study liver vascular lesions were constantly seen in the chronic phase of FE near affected bile ducts. It is known that extrahepatic and intrahepatic bile ducts are located with branches of the hepatic artery (their sole blood supply) and portal vein [39]. Eccentric subintimal fibroblastic proliferation on the side adjacent to affected bile ducts were sometimes seen in sheep and goats [1,8] and adult cows [24]. It has been suggested that a high concentration of sporidesmin injures the biliary epithelium and the release of toxin and bile acids produce irritative lesions and coagulative necrosis of blood vessels, both arteries and veins [34].

Ductular reaction (DR) appears to be one of the factors that deteriorate liver function, because gradually replaces the hepatic parenchyma and causes a gradual decrease in mature hepatocytes [40]. In this study DR was recognized as bile duct hyperplasia in extensive areas of the liver in chronic FE cases and the cells present in the lesion, immunostained positively for pancytokeratins but did not express mesenchymal cell markers [23]. DR is observed in cholestatic liver diseases and is closely related to liver fibrosis induced by HSCs and portal fibroblasts and is also an important factor for liver regeneration [41]. According to these last authors, the mechanism responsible for DR is not definitively understood and cholangiocytes, hepatocytes, or hepatic progenitor cells can be the origin of active cells during DR, depending on specific liver injury.

In FE cirrhotic livers the fibrous septa contain large numbers of HSCs expressing protein α-SMA. It is known that, as HSCs activate, the expression of α-SMA is increased, which confers contractile potential to the cells [25]. HSCs are the major source of type I collagen and other extracellular matrix proteins that characterize the fibrotic liver [25,42].

The sporidesmin elicits biliary insult and an activation of HSCs alongside induction of hepatic inflammation. In the acute exposure to this toxin, an inflammatory response of neutrophils, lysozyme and calprotectin in KCs was observed in areas of cholestasis. This fact could suggest that KC response occurs early in cholestatic injury and bile acids leakage from cholangioles may be involved in this proliferation [27]. Previous studies indicate that HSCs activation also promotes the recruitment of leukocytes in the early phase of liver injury [26]. In chronic FE these macrophages were seen mainly around degenerated bile ducts and, to a lesser extent, in relation to DR. Nevertheless CD206+ macrophages were more numerous in this last location. These results suggest that in the acute and chronic phases of FE, the peribiliar inflammatory infiltrate is dominated by classically activated M1 macrophages. As a detail, in the chronic phase there was an increase of activated M2 macrophages associated with increased fibrogenesis [43]. These data indicated that M1 macrophages prevail during the onset of liver injury, and M2 macrophages, if liver injury becomes chronic, take up a profibrotic role secreting TGF-β. In this work TGF-β expression was observed in cells consistent with macrophages in fibrotic septa in areas of DR. There is evidence supporting an indirect role of M2 macrophages in the development of fibrosis secreting factors like TGF-β, which activate HSCs [44,45]. Besides this, in the present work, TGF-β immunostaining was observed in hepatic arterial vessels with vascular occlusive lesion. This cytokine plays a prominent role in vascular disorders such as the arterial thickening associated with pulmonary hypertension [46] and in other arterial pathologies, with effects on the changes of the vascular smooth-muscle cells during transition from structural to a synthetic phenotype [47]. It has been also documented that both M1 and M2 macrophages accumulate in fibrotic septa of mouse and human end-stage cirrhotic livers, suggesting that both are necessary in fibrotic responses [45]. An interesting finding in the hepatic immune response in the case of FE was the presence of numerous T CD3+ lymphocytes in proximity to HSCs in the fibrous septa and around damaged bile ducts in chronic FE lesions. There is evidence that HSCs secrete cytokines, such as TGF-β, that has lymphocyte chemotactic activity and contributes to recruit and positioning of lymphocytes within the liver stroma in order to maintain an effective immune response [26]. In addition, it has been suggested that T lymphocytes can interact with HSCs and secrete various cytokines to modulate and sustaining fibrotic responses in chronic liver disease [26,48]. In chronic FE sheep an accumulation of CD3+ lymphocytes, and few plasma cells expressing IgG were observed, similarly to alpha-naphthylisothiocyanate (ANIT)-induced biliary pathology in mice and other cholestatic liver diseases in humans [49]. The vascular and ductal system and its macroscopic shape of sheep's liver are very similar to the human organ, so it has a great potential as an animal model [50].

5. Conclusions

FE eczema causes loss of weight, skin lesions and liver damage principally. The biochemical parameters of the blood samples suggest that GGT and AST are elevated in animals affected. Lesions such as liver atrophy and cirrhosis are characterized by a ductular reaction with the associated presence of HSCs, KCs, lymphocytes and neutrophils. These features vary depending on the time of previous exposure to the toxin, with a pro-inflammatory character in the acute phases and an anti-inflammatory, and therefore profibrotic, component in the chronic cases. T-lymphocyte infiltrates persist in chronic forms, in association with increased levels of HSCs, which could contribute to recruit lymphocytes within the liver stroma in order to maintain an effective immune response. Considering that the vascular and ductal system of the ovine liver have strong similarities to the human organ, sheep have remarkable potential as a suitable animal model to study the pathogenesis and the interaction of toxicosis that result in liver damage.

Author Contributions: Design and supervision, M.C.F.; design and writing the paper M.F. (Miguel Fernández) and M.C.F.; draft preparation, M.F. (Miguel Fuertes); sample collection and histopathological analysis, M.F. (Miguel Fernández), M.F. (Miguel Fuertes), V.P. and M.C.F.; plot and clinical information, J.M.; mycological study A.L.G.-P.; statistical analysis J.B. and J.E. All authors have read and agreed to the published version of the manuscript.

Funding: This work was supported by the research funds of the Ruminant Health and Pathology research group of the University of León.

Institutional Review Board Statement: Ethical review and approval were waived for this study, due to the type of study in which the blood samples were obtained during regular veterinary clinical services and with post mortem tissue samples that are routinely collected after the death of animals

Data Availability Statement: Supporting data is available to applicants through the correspondence address.

Acknowledgments: The authors wish to thank Antonio Martínez for providing the animals for this study and Marta Silva for technical assistance.

Conflicts of Interest: None of the authors has any financial or personal relationships that could inappropriately influence or bias the content of the paper.

References

1. Smith, B.L.; Embling, P.P. Facial eczema in goats: The toxicity of sporidesmin in goats and its pathology. *N. Z. Vet. J.* **1991**, *39*, 18–22. [CrossRef] [PubMed]
2. Gardiner, D.M.; Waring, P.; Howlett, B. The epipolythiodioxopiperazine (ETP) class of fungal toxins: Distribution, mode of action, functions and biosynthesis. *Microbiology* **2005**, *151*, 1021–1032. [CrossRef] [PubMed]
3. Riet-Correa, F.; Rivero, R.; Odriozola, E.; Adrien, M.D.L.; Medeiros, R.M.T.; Schild, A.L. Mycotoxicoses of ruminants and horses. *J. Vet. Diagn. Investig.* **2013**, *25*, 692–708. [CrossRef] [PubMed]
4. Worker, N.A. A hepatotoxin causing liver damage in facial eczema of sheep. *Nature* **1960**, *185*, 909–910. [CrossRef] [PubMed]
5. Cheeke, P.R. Endogenous toxins and mycotoxins in forage grasses and their effects on livestock. *J. Anim. Sci.* **1995**, *73*, 909–918. [CrossRef]
6. Smith, B.L.; Towers, N.R. Mycotoxicoses of grazing animals in New Zealand. *N. Z. Vet. J.* **2002**, *50*, 28–34. [CrossRef]
7. Di Menna, M.E.; Smith, B.L.; Miles, C.O. A history of facial eczema (pithomycotoxicosis) research. *N. Z. J. Agric. Res.* **2009**, *52*, 345–376. [CrossRef]
8. Marasas, W.F.; Adelaar, T.F.; Kellerman, T.S.; A Minné, J.; Van Rensburg, I.B.; Burroughs, G.W. First report of facial eczema in sheep in South Africa. *Onderstepoort J. Vet. Res.* **1972**, *39*, 107–112.
9. Edwards, J.R.; Richards, R.B.; Love, R.A.; Harrison, M.R.; Gwynn, R.V.R. An abattoir survey of the prevalence of facial eczema in sheep in Western Australia. *Aust. Vet. J.* **1983**, *60*, 157. [CrossRef]
10. Hansen, D.E.; McCoy, R.D.; Hedstrom, O.R.; Snyder, S.P.; Ballerstedt, P.B. Photosensitization associated with exposure to Pithomyces chartarum in lambs. *J. Am. Vet. Med. Assoc.* **1994**, *204*, 1668–1671.
11. Bezille, P.; Braun, J.P.; le Bars, J. First identification of facial eczema in ewes in Europe. *Recl. Méd. Vét.* **1984**, *160*, 339–347.
12. Pinto, C.; Santos, V.M.; Dinis, J.; Peleteiro, M.C.; Fitzgerald, J.M.; Hawkes, A.D.; Smith, B.L. Pithomycotoxicosis (facial eczema) in ruminants in the Azores, Portugal. *Vet. Rec.* **2005**, *157*, 805–810. [CrossRef]
13. Van Wuijckhuise, L.; Snoep, J.; Cremers, G.; Duvivier, A.; Groeneveld, A.; Ottens, W.; Van Der Sar, S. First case of pithomycotoxicosis (facial eczema) in the Netherlands. *Tijdschr. Diergeneeskd.* **2006**, *131*, 858–861.
14. Ozmen, O.; Sahinduran, S.; Haligur, M.; Albay, M.K. Clinicopathological studies on facial eczema outbreak in sheep in Southwest Turkey. *Trop. Anim. Health Prod.* **2008**, *40*, 545–551. [CrossRef]
15. González, L.; Marco, J.C.; Oregui, L.M.; Cuervo, L.; Korkostegi, J.L. Observaciones sobre el eczema facial de las ovejas. In *ITEA, Vol Extra, nº 11, Tomo, II*; Ed Alabart, J.L., Alberti, P., Blasco, A., García-Marín, J.F., Manrique, E., Purroy, A., Eds.; Editorial Alabart: Zaragoza, Spain, 1991; pp. 674–676.
16. Ford, E.J.H. Activity of gamma-glutamyl transpeptidase and other enzymes in the serum of sheep with liver or kidney damage. *J. Comp. Pathol.* **1974**, *84*, 231–243. [CrossRef]
17. Thompson, K.G.; Jones, D.H.; Sutherland, R.J.; Camp, B.J. Sporidesmin toxicity in rabbits: Biochemical and morphological changes. *J. Comp. Pathol.* **1983**, *93*, 319–329. [CrossRef]
18. Bonnefoi, M.; Braun, J.P.; Bézille, P.; LeBars, J.; Sawadogo, G.; Thouvenot, J.P. Clinical biochemistry of sporidesmin natural intoxication (facial eczema) of sheep. *J. Clin. Chem. Clin. Biochem.* **1989**, *27*, 13–18. [CrossRef]

19. Towers, N.R.; Stratton, G.C. Serum gamma-glutamyltransferase as a measure of sporidesmin-induced liver damage in sheep. *N. Z. Vet. J.* **1978**, *26*, 109–112. [CrossRef]
20. Smith, B.L. Effects of low dose rates of sporidesmin administered orally to sheep. *N. Z. Vet. J.* **2000**, *48*, 176–181. [CrossRef]
21. Smith, B.L.; Embling, P.P.; Gravett, I.M. Pithomyces chartarum spore counts in rumen contents and faeces of sheep exposed to autumn pasture at three different grazing pressures. *J. Appl. Toxicol.* **1987**, *7*, 179–184. [CrossRef]
22. Strazzabosco, M.; Fiorotto, R.; Cadamuro, M.; Spirli, C.; Mariotti, V.; Kaffe, E.; Scirpo, R.; Fabris, L. Pathophysiologic implications of innate immunity and autoinflammation in the biliary epithelium. *Biochim. Biophys. Acta Mol. Basis Dis.* **2018**, *1864*, 1374–1379. [CrossRef] [PubMed]
23. Lee, S.J.; Park, J.B.; Kim, K.H.; Lee, W.R.; Kim, J.Y.; An, H.J.; Park, K.K. Immunohistochemical study for the origin of ductular reaction in chronic liver disease. *Int. J. Clin. Exp. Pathol.* **2014**, *7*, 4076–4085. [PubMed]
24. Collett, M.G. Bile duct lesions associated with turnip (Brassica rapa) photosensitization compared with those due to sporidesmin toxicosis in dairy cows. *Vet. Pathol.* **2014**, *51*, 986–991. [CrossRef] [PubMed]
25. Friedman, S.L. Hepatic fibrosis-Overview. *Toxicology* **2008**, *254*, 120–129. [CrossRef]
26. Holt, A.P.; Salmon, M.; Buckley, C.D.; Adams, D.H. Immune interactions in hepatic fibrosis or "leucocyte-stromal interactions in hepatic fibrosis". *Clin. Liver Dis.* **2008**, *12*, 861–882. [CrossRef]
27. Hines, J.E.; Johnson, S.J.; Burt, A.D. In vivo responses of macrophages and perisinusoidal cells to cholestatic liver injury. *Am. J. Pathol.* **1993**, *142*, 511–518.
28. CES. Metereología y climatología. In *Estado del Medio Ambiente en Asturias 2003*; Consejo Económico y Social (CES) del Principado de Asturias: Oviedo, Spain, 2005; pp. 27–46.
29. Aiello, S.E.; Mays, A.; Amstuz, H.E.; Anderson, D.P.; Armour, J. *El Manual Merck de Veterinaria*, 5th ed.; Océano Grupo Editorial SA: Barcelona, Spain, 2000; pp. 2454–2455.
30. Kellerman, T.S.; Van der Westhuizen, G.C.; Coetzer, J.A.; Roux, C.; Marasas, W.F.; Minne, J.A.; Bath, G.F.; Basson, P.A. Photosensitivity in South Africa. II. The experimental production of the ovine hepatogenous photosensitivity disease geeldikkop (Tribulosis ovis) by the simultaneous ingestion of Tribulus terrestris plants and cultures of Pithomyces chartarum containing the mycotoxin sporidesmin. *Onderstepoort J. Vet. Res.* **1980**, *47*, 231–261.
31. Lindsay, G.C.; Morris, C.A.; Boucher, M.; Capundan, K.; Jordan, T.W. Effects of sporidesmin on cultured biliary tract cells from Romney lambs that differed in their sensitivity to sporidesmin. *N. Z. Vet. J.* **2018**, *66*, 325–331. [CrossRef]
32. Gollan, J.L.; Billing, B.H.; Huang, S.N. Ultrastructural changes in the isolated rat kidney induced by conjugated bilirubin and bile acids. *Br. J. Exp. Pathol.* **1976**, *57*, 571–581.
33. Van Slambrouck, C.M.; Salem, F.; Meehan, S.M.; Chang, A. Bile cast nephropathy is a common pathologic finding for kidney injury associated with severe liver dysfunction. *Kidney Int.* **2013**, *84*, 192–197. [CrossRef]
34. Cullen, J.M.; Stalker, M.J. Liver and biliary system. In *Jubb, Kennedy, and Palmer's Pathology of Domestic Animals*, 6th ed.; Maxie, M.G., Ed.; Elsevier: St. Louis, MO, USA, 2016; pp. 258–352.
35. Tseng, C.A.; Pan, Y.S.; Chen, C.Y.; Liu, C.S.; Wu, D.C.; Wang, W.M.; Jan, C.M. Biliary cystadenocarcinoma associated with atrophy of the left hepatic lobe and hepatolithiasis mimicking intrahepatic cholangiocarcinoma: A case report. *Kaohsiung J. Med. Sci.* **2004**, *20*, 198–203. [CrossRef]
36. Isobe, K.; Nakayama, H.; Uetsuka, K. Relation between lipogranuloma formation and fibrosis, and the origin of brown pigments in lipogranuloma of the canine liver. *Comp. Hepatol.* **2008**, *7*, 5. [CrossRef]
37. Munday, R. Studies on the mechanism of toxicity of the mycotoxin sporidesmin. IV Inhibition by copper-chelating agents of the generation of superoxide radical by sporidesmin. *J. Appl. Toxicol.* **1985**, *5*, 69–73. [CrossRef]
38. López Panqueva, R.P. Useful algorithms for histopathological diagnosis of liver disease based on patterns of liver damage. *Rev. Colomb. Gastroenterol.* **2016**, *31*, 436–449.
39. Babu, C.S.R.; Sharma, M. Biliary tract anatomy and its relationship with venous drainage. *J. Clin. Exp. Hepatol.* **2014**, *4*, 518–526.
40. Nishikawa, Y.; Sone, M.; Nagahama, Y.; Kumagai, E.; Doi, Y.; Omori, Y.; Yoshioka, T.; Tokairin, T.; Yoshida, M.; Yamamoto, Y.; et al. Tumor necrosis factor-α promotes bile ductular transdifferentiation of mature rat hepatocytes in vitro. *J. Cell. Biochem.* **2013**, *114*, 831–843. [CrossRef]
41. Sato, K.; Marzioni, M.; Meng, F.; Francis, H.; Glaser, S.; Alpini, G. Ductular reaction in liver diseases: Pathological mechanisms and translational significances. *Hepatology* **2019**, *69*, 420–430. [CrossRef]
42. Gazdic, M.; Arsenijevic, A.; Markovic, B.S.; Volarevic, A.; Dimova, I.; Djonov, V.; Arsenijevic, N.; Stojkovic, M.; Volarevic, V. Mesenchymal stem cell-dependent modulation of liver diseases. *Int. J. Biol. Sci.* **2017**, *13*, 1109–1117. [CrossRef]
43. Fabregat, I.; Caballero-Díaz, D. Transforming growth factor-β-induced cell plasticity in liver fibrosis and hepatocarcinogenesis. *Front. Oncol.* **2018**, *8*, 357. [CrossRef]
44. Baldus, S.E.; Zirbes, T.K.; Weidner, I.C.; Flucke, U.; Dittmar, E.; Thiele, J.; Dienes, H.P. Comparative quantitative analysis of macrophage populations defined by CD68 and carbohydrate antigens in normal and pathologically altered human liver tissue. *Anal. Cell. Pathol.* **1998**, *16*, 141–150. [CrossRef]
45. Beljaars, L.; Schippers, M.; Reker-Smit, C.; Martinez, F.O.; Helming, L.; Poelstra, K.; Melgert, B.N. Hepatic localization of macrophage phenotypes during fibrogenesis and resolution of fibrosis in mice and humans. *Front. Immunol.* **2014**, *5*, 430. [CrossRef]

46. Roberts, A.B.; Sporn, M.B. Regulation of endothelial cell growth, architecture, and matrix synthesis by TGF-beta. *Am. Rev. Respir. Dis.* **1989**, *140*, 126–128. [CrossRef] [PubMed]
47. Lyons, J.A.; Dickson, P.I.; Wall, J.S.; Passage, M.B.; Ellinwood, N.M.; Kakkis, E.D.; McEntee, M.F. Arterial pathology in canine mucopolysaccharidosis-I and response therapy. *Lab. Investig.* **2011**, *91*, 665–674. [CrossRef] [PubMed]
48. Xing, Z.Z.; Huang, L.Y.; Wu, C.R.; You, H.; Ma, H.; Jia, J.D. Activated rat hepatic stellate cells influence Th1/Th2 profile in vitro. *World J. Gastroenterol.* **2015**, *21*, 7165–7171. [CrossRef] [PubMed]
49. Joshi, N.; Kopec, A.K.; Cline-fedewa, H.; Luyendyk, J.P. Lymphocytes contribute to biliary injury and fibrosis in experimental xenobiotic-induced cholestasis. *Toxicology* **2017**, *377*, 73–80. [CrossRef]
50. Venkatesh, B.; Purushotham, G.; Pramod Kumar, D.; Raghavender, K.B.P. Anatomical distribution of the hepatic artery in sheep. *J. Pharmacogn. Phytochem.* **2018**, *SP1*, 2529–2533.

Case Report

Dentinogenic Ghost Cell Tumor in a Sumatran Rhinoceros

Annas Salleh [1,*], Zainal Z. Zainuddin [2], Reza M. M. Tarmizi [2], Chee K. Yap [2], Chian-Ren Jeng [3] and Mohd Zamri-Saad [1]

1   Department of Veterinary Laboratory Diagnosis, Faculty of Veterinary Medicine, Universiti Putra Malaysia, Serdang 43400, Selangor, Malaysia; mzamri@upm.edu.my

2   Borneo Rhino Alliance, c/o Faculty of Sciences and Natural Resources, Universiti Malaysia Sabah, Kota Kinabalu 88400, Sabah, Malaysia; zainalz.bora@gmail.com (Z.Z.Z.); reza2727@gmail.com (R.M.M.T.); kcyap@upm.edu.my (C.K.Y.)

3   Graduate Institute of Molecular and Comparative Pathobiology, School of Veterinary Medicine, National Taiwan University, Taipei 106216, Taiwan; crjeng@ntu.edu.tw

*   Correspondence: annas@up.edu.my

Simple Summary: A dentinogenic ghost cell tumor is an odontogenic ghost cell lesion of the maxilla and mandible. It is a rare tumor that has been described in humans. This work describes the clinical and pathological findings of an advanced stage of a dentinogenic ghost cell tumor, a type that has not previously been described in veterinary medicine. The advanced stage of this tumor led to the observation of aberrant keratinization, characterized by ghost cells and numerous islands of dentinoid formation. Diagnosis was made with the aid of routine histology, special histochemistry, immunohistochemistry, and classification and features from human oncology as a reference.

Abstract: An adult female Sumatran rhinoceros was observed with a swelling in the left infraorbital region in March 2017. The swelling rapidly grew into a mass. A radiograph revealed a cystic radiolucent area in the left maxilla. In June 2017, the rhinoceros was euthanized. At necropsy, the infraorbital mass measured 21 cm × 30 cm. Samples of the infraorbital mass, left parotid gland, and left masseter muscle were collected for histopathology (Hematoxylin & Eosin, Von Kossa, Masson's trichrome, cytokeratin AE1/AE3, EMA, p53, and S-100). Numerous neoplastic epithelial cells showing pleomorphism and infiltration were observed. Islands of dentinoid material containing ghost cells and keratin pearls were observed with the aid of the two special histochemistry stains. Mitotic figures were rarely observed. All the neoplastic odontogenic cells and keratin pearls showed an intense positive stain for cytokeratin AE1/AE3, while some keratin pearls showed mild positive stains for S-100. All samples were negative for p53 and S-100 immunodetection. The mass was diagnosed as a dentinogenic ghost cell tumor.

Keywords: dentinogenic ghost cell tumor; odontogenic ghost cell lesion; *Sumatran rhinoceros*; *Dicerorhinus sumatrensis*; immunohistochemistry; special stain

Citation: Salleh, A.; Zainuddin, Z.Z.; Tarmizi, R.M.M.; Yap, C.K.; Jeng, C.-R.; Zamri-Saad, M. Dentinogenic Ghost Cell Tumor in a Sumatran Rhinoceros. *Animals* **2021**, *11*, 1173. https://doi.org/10.3390/ani11041173

Academic Editors: Alejandro Suárez-Bonnet and Gustavo A. Ramírez Rivero

Received: 29 January 2021
Accepted: 16 April 2021
Published: 20 April 2021

1. Introduction

In humans, a few types of tumors are identified as odontogenic ghost cell lesions (OGCL) of the maxilla and mandible. This includes calcifying odontogenic cysts (COC), dentinogenic ghost cell tumors (DGCT), and ghost cell odontogenic carcinoma (GCOC) [1]. DGCT is a benign but locally infiltrative neoplasm of odontogenic epithelium. It is a rare tumor in humans with very limited reports. A ghost cell is an enlarged epithelial cell having an eosinophilic cytoplasm with a faint nucleus outline or no nucleus [2]. It is associated with a marked aberrant keratinization. DGCT has been described as a rare form of ghost cell lesion, accounting for 3–5% of all cases involving ghost cell lesions [3,4].

For OGCL in humans, the prognosis and recurrence rate may differ according to the type of tumor. For COC, prognosis is considered excellent and the recurrence rate is low. When recurrence of COC occurs, it typically involves elderly persons [5]. Recurrence in young persons is rarely reported [6]. For DGCT, reports on recurrence rates range between 33% and 73% [7]. Surgical removal of DGCT involving an extensive procedure usually results in a low recurrence rate, while simple enucleation of the tumor usually results in a higher recurrence rate. Recurrence may occur within 5 to 10 years [8]. GCOC has a 73% five-year survival rate, and recurrence is reported to be common [9].

To our knowledge, DGCT has never been documented in animals. This article reports the first case of DGCT in an animal.

2. Description of the Case

A female Sumatran rhinoceros (*Dicerorrhinus sumatrensis*) weighing 508 kg and estimated to be between 25 and 30 years old was managed in a one-hectare forested paddock at the Tabin Wildlife Reserve, Sabah, Malaysia. In January 2017, it showed signs of difficulty in mastication, especially chewing on larger stems. Subsequently, in February 2017, it developed a 5 cm left unilateral, infraorbital and maxillary swelling with epiphora. It was treated with oral flunixin meglumine (Banamine® at 1500 mg per day for 3 days, and oral amoxicillin and clavulanate potassium (Augmentin™) for five consecutive days. However, within a month, the swelling rapidly developed into a firm mass measuring about 15 cm in diameter, which later ruptured to discharge a mucopurulent exudate. In addition to wound cleaning twice a day, the rhinoceros was treated with oral amoxicillin and clavulanate potassium (Augmentin™) at 25 mg/kg for 5 days, and parenteral dexamethasone (Dexadreson®) at 0.1 mg/kg intramuscularly for 3 days. Despite the treatment, the wound did not show any improvement and eventually became a 5 cm over-granulated open wound with blood-tinged nasal discharge from the left nostril. At this point, the appetite and body weight were slightly reduced, while the right jaw was predominantly used for mastication.

Staphylococcus sp. was isolated from the swab sample of the open wound, while an antibiotic sensitivity test showed resistance to amoxicillin-clavulanate acid but susceptibility to enrofloxacin and cephalosporin. In April 2017, a radiograph revealed a unilocular radiolucent area surrounding the 2nd and 3rd maxillary cheek teeth, suggestive of a cyst (Figure 1A). This cyst was connected to the paranasal sinuses by an oronasal fistula. A radiopaque fragment was noted dorsal to the 3rd maxillary cheek tooth, indicating a fracture of alveolar bone. The rhinoceros was orally treated with dexamethasone, Augmentin™, lactated Ringer's solution, dextrose, Duphalyte, vitamin K, iron supplement, and phenylbutazone.

Dental extraction surgery was performed with peri-operative treatment comprising flunixin meglumine and enrofloxacin. Three cheek teeth (1st, 2nd, and 3rd cheek teeth) were successfully extracted in the surgery. All the extracted teeth had yellowish expansile solid masses around the roots. However, the oronasal fistula was not examined, as it could not be reached through the alveolar opening. For post-operative treatment, phenylbutazone, enrofloxacin, ceftiofur, and oral rinse were administered.

Thirty minutes after the recovery from anesthesia, the animal regained normal appetite. Wound cleaning, mouth wash, and parenteral enrofloxacin once daily, every other day were continued. However, the open wound, nasal discharge, and epiphora persisted. Thus, the antibiotic was changed to ceftiofur on day 8 after dental extraction. The bodyweight increased to 512 kg 7 days after the surgery. The intraoral granulation tissue eventually subsided. However, between May and June 2017, the animal showed occasional epistaxis and dyspnea, while the cutaneous mass aggressively grew larger. The rhinoceros was euthanized by intravenous administration of detomidine, ketamine, and pentobarbitone.

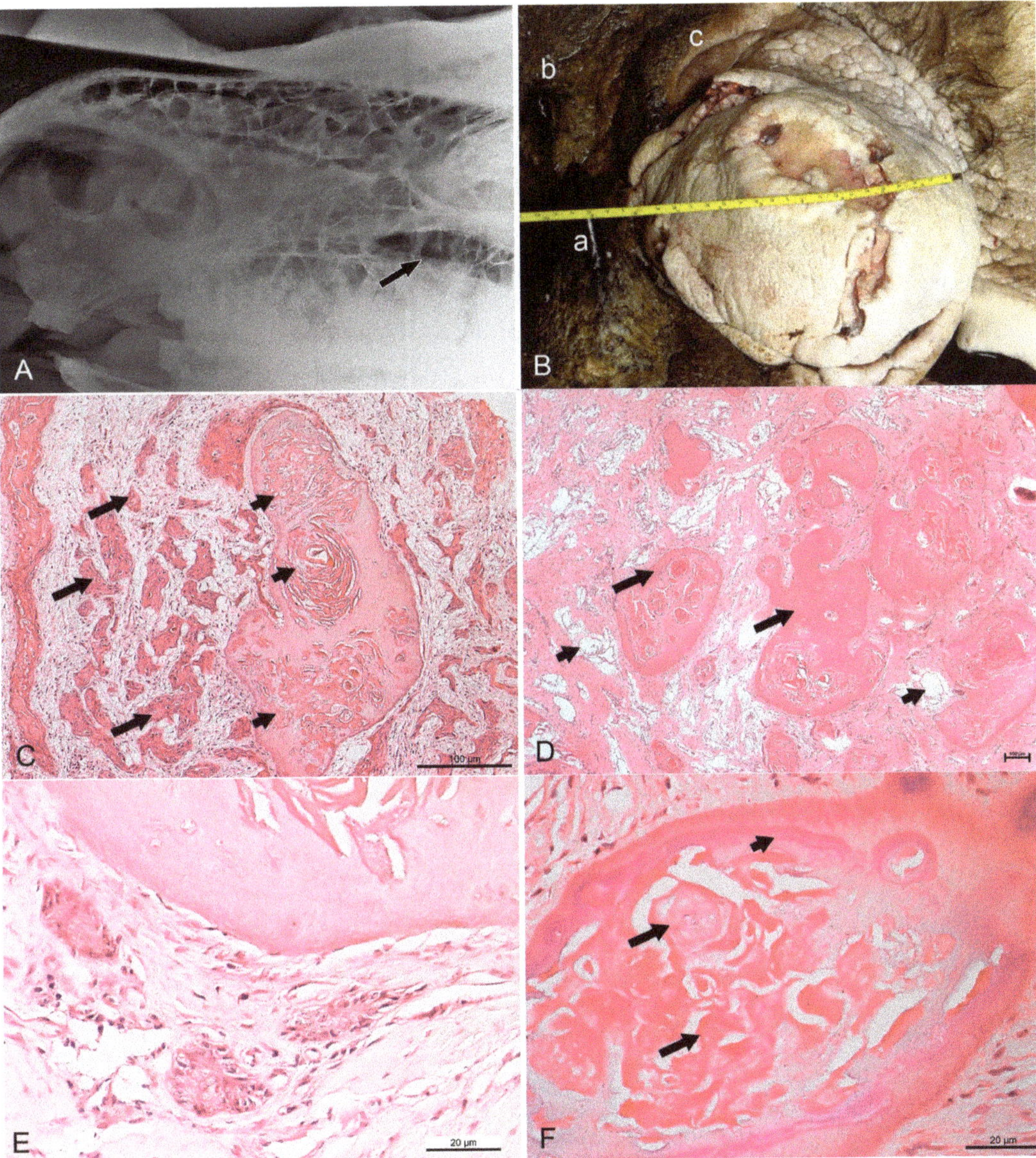

Figure 1. Radiographic, gross, and routine histopathological findings in a Sumatran rhinoceros with a dentinogenic ghost cell tumor. (**A**) Left-lateral view radiograph taken in April 2017 showing a unilocular radiolucent cyst (arrow) at the left maxilla. (**B**) The infraorbital mass in June 2017 measuring 21 cm × 30 cm. a: anterior horn, b: posterior horn, c: left upper eyelid. (**C**) Nests of neoplastic squamous cells (long arrows) surrounded by substantial compact fibrous stroma. Note the formation of keratin pearls (short arrows) embedded in a dentinoid material. HE (hematoxylin and eosin), bar = 100 μm. (**D**) Numerous islands of dentinoid material (long arrows) surrounded by loose and vascularized stroma (short arrows). HE, bar = 100 μm. (**E**) Neoplastic cells showing pleomorphism with basaloid- or stellate-reticulum-like appearance arranged in a nest. Vesicular nuclei can be observed. HE, bar = 20 μm (**F**) Ghost cell (long arrows) at the center of dentinoid material with cementum-like appearance (short arrow). HE, bar = 20 μm.

During the post-mortem examination, a tissue mass was visible around the dental extraction site, with the remaining 2nd and 3rd molars having enormous amounts of the expansile solid mass around the crowns and roots. The skin around the open wound was edematous and swollen. The infraorbital mass measured 21 × 30 cm (Figure 1B), with several open wounds of 1 to 7 cm in diameter. The mass extended ventrally and dorsal into the eyes. The size and color of the left masseter muscles were darker compared to the opposite side, suggestive of degenerative changes. Fistula between the maxilla and infraorbital mass was noted, while the left parotid gland was gritty with whitish spots. No metastasis to either adjacent or distant organs was observed. Samples from the infraorbital mass, left parotid gland, and left masseter muscle were collected and fixed in 10% neutral-buffered formalin, routinely processed, and stained with hematoxylin and eosin (HE), special histochemical Masson's trichrome and Von Kossa stains, and immunohistochemistry was conducted for detection of cytokeratin AE1/AE3, epithelial membrane antigen (EMA), p53, and S-100.

In the infraorbital mass, islands of neoplastic epithelium of various sizes were observed embedded or infiltrated in substantial amounts of either compact or loose fibrous stroma (Figure 1C). In some areas, the stroma was extensively loose with increased vascularization (Figure 1D). The neoplastic cells showed an infiltrative growth pattern arranged in strands, unsuccessful anastomosing, or medusa-like patterns. A long trabecular arrangement of tumor cells was observed. Multifocal squamous metaplasia or keratin-like material deposition was noted in the centers of the tumor islands. The tumor cells could be seen surrounding and embedded in numerous islands of dentinoid material. In addition, accumulation of pale eosinophilic ghost cells and a whirl-like arrangement of keratin-like material infiltrating the dentinoid material were noticeable. At high magnification, the neoplastic cells showed pleomorphism with a basaloid- or stellate-reticulum-like appearance with vesicular nuclei, usually arranged in a nest (Figure 1E). Ghost cells, characterized by large, eosinophilic cells that contained either the outline of a nucleus or no nucleus, were present at the cementum-like appearance of the dentinoid materials (Figure 1F). Mitoses were occasionally seen. The mitotic count, determined using a previously described method, was a low count of 2 [10]. The dentinoid was further confirmed by positive staining using Von Kossa stain to indicate the presence of calcium, and blue staining by Masson's trichrome stain. Most of the keratin and ghost cells lacked calcium, as indicated by the negative staining by Von Kossa stain (Figure 2A) and red staining by Masson's trichrome stain (Figure 2B). Some keratin pearls were observed without dentinoid formation, but they were surrounded by substantial amounts of neoplastic epithelial cells. The left masseter muscle was mildly degenerated but showed no evidence of invasion by neoplastic cells, while the left parotid gland was severely calcified.

The neoplastic epithelial cells showed intense intracytoplasmic immunodetection of cytokeratin AE1/AE3 but were negative for EMA and S-100. All keratin pearls, including those found inside the dentinoid material, and most of the ghost cells, showed intense staining with cytokeratin AE1/AE3 (Figure 2C), mild staining against S-100 (Figure 2D), and negative against p53 and EMA.

The differential diagnoses for this case included ghost cell odontogenic carcinoma (GCOC), dentinogenic ghost cell tumor (DGCT), craniopharyngioma, primary intraosseous squamous cell carcinoma (PIOSCC), squamous cell carcinoma (SCC), and ameloblastoma. The radiology and histopathology examinations established the diagnosis of DGCT.

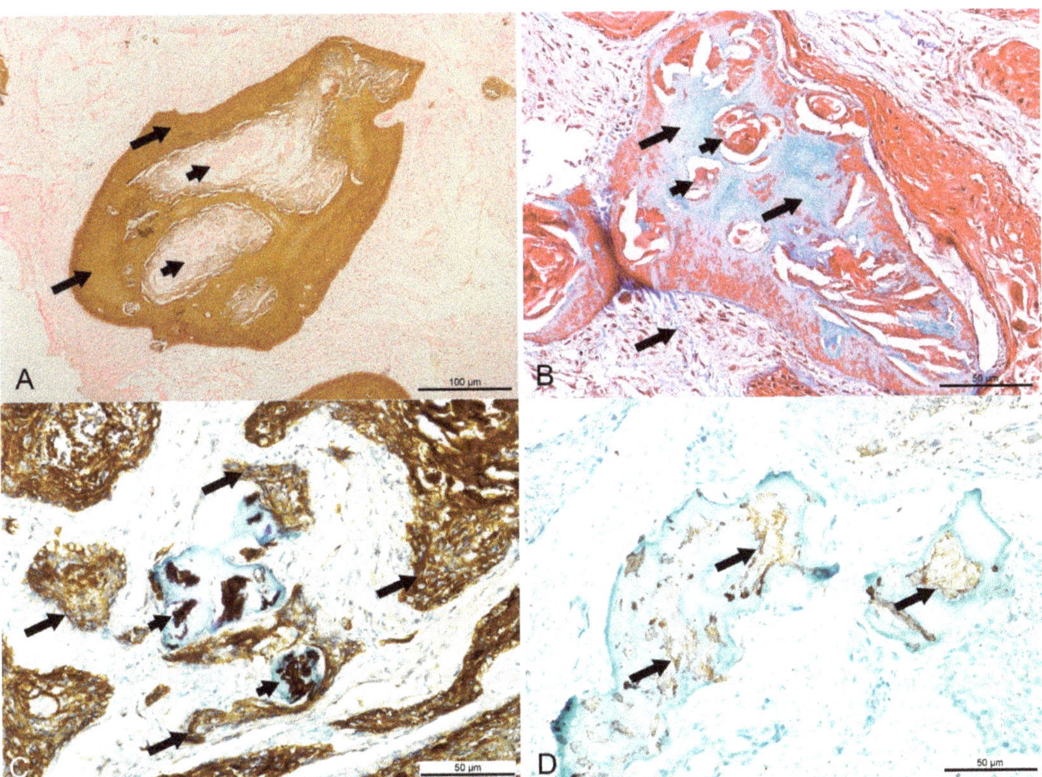

Figure 2. Special histochemical and immunohistochemistry findings in a Sumatran rhinoceros with a dentinogenic ghost cell tumor. (**A**) Brown stain of Von Kossa indicating the presence of calcium in the dentinoid material (long arrow), while most keratin pearls were devoid of calcium (short arrows). Von Kossa, bar = 100 μm. (**B**) Bone tissue stained in blue and keratin stained in red with Masson's trichrome. Masson's trichrome, bar = 50 μm. (**C**) Intense intracytoplasmic staining for cytokeratin AE1/AE3 in the neoplastic squamous cells (long arrows) and keratin pearls (short arrows). AE1/AE3, bar = 50 μm. (**D**) Mild positive staining for S-100 in the ghost cells and keratin (arrows) located inside the dentinoid material. S-100, bar = 50 μm.

3. Discussion

OGCL are considered challenging to diagnose, as COC, DGCT, and GCOC have similar histological features [11]. Diagnosis of DGCT in this rhinoceros was largely made based on the histological and immunohistochemical features from human oncology and pathology as compiled in Table 1. From the differential diagnoses, PIOSCC, SCC and ameloblastoma were ruled out, as these tumors do not feature ghost cell lesions [12]. GCOC was ruled out mainly by the fact that histopathological examination showed low mitotic activity, suggestive of a benign cellular status. Furthermore, it did not invade adjacent tissues, lacked necrosis, had pleomorphic neoplastic cells, and the immunohistochemistry for p53 was negative. Although about 30% of GCOC may show negativity for p53, it has been reported that the diagnosis of GCOC versus DGCT should be largely based on p53 positivity [3,13]. Formation of dentinoid material in craniopharyngioma is extremely rare. If present, these dentinoid materials are described as not obvious [14], so craniopharyngioma was ruled out. From this case and bibliographical review, it was stated that differentiating DGCT, GCOC, and other differential diagnoses based on the epithelial histological and immunochemical features can be difficult. The reason for this is that they may show similar epithelial features ranging from palisading columnar (resembling ameloblastoma) to basaloid (resembling squamous epithelium) formations. The presence of foreign body giant cells has been reported in both DGCT and GCOC [15,16]. Because of the rarity of OGCL and the many synonyms for each OGCL neoplasm, available data pertaining to their immunohistochemical characteristics may be difficult to access.

Table 1. Summary of histological and immunohistochemical features of COC, DGCT, and GCOC in humans.

	COC	DGCT	GCOC	References
		Histological Features		
Cyst component	Main	Occasional	Occasional	[17]
Epithelium	Mainly cystic Palisading columnar cells resembling ameloblastoma	Tumorous and occasionally cystic Ameloblastous or basaloid	Tumorous and rarely cystic Uniform small basaloid, with round or vesicular nuclei	[3,17]
Ghost Cell	Consistent	Marked	Predominant	[17]
Calcification	Frequent	Occasional	Rare	[17]
Dentinoid Material	None	Predominant	Rudimentary	[17]
Cellular status	Benign	Benign	Malignant	[17]
Mitosis	Present	Rare	Frequent	[3]
Recurrence	Rare	Rare	Frequent	[17]
		Immunohistochemistry Features		
Cytokeratin AE1/AE3	+	+	+	[18,19]
Beta catenin	+	+	+	[20,21]
S-100	+/−	+	+/−	[18,22–24]
EMA	n/a	n/a	−	[18]
p53	n/a	+/−	+/− (>70% of cases show +)	[3,18,19,25]

In general, DGCT more commonly occurs in the posterior maxilla and mandible. A slight predilection for the mandible has been reported, where 53% of DGCT occurs in the mandible [3]. Two variants of DGCT, namely, central and peripheral, have been described [26]. Central DGCT, the more common of the two, is a locally invasive intraosseous tumor, whereas peripheral DGCT is a non-invasive extraosseous tumor [27]. In most cases of central DGCT, the radiographic features are unilocular with a mixture of radiolucent and radiopaque or only radiolucent lesions [3]. It is unfortunate that no sample from the maxillary cyst was collected and examined in this case. The radiographic observation of mandibular cyst in this rhinoceros suggested that this case involved a central DGCT.

Cases of other OGCL in animals have been previously reported, such as epithelial ghost cells and dentinoid material in rats with odontogenic tumors [28]. Another report involved a Bengal tiger (*Panthera tigris tigris*), wherein only a few ghost cells and some keratin were observed in a mandibular mass. However, no formation of dentinoid material was observed. That case was diagnosed as calcifying epithelial odontogenic tumor [29]. It is possible that the lack of reports of GCOC is due to the rarity of the tumor or the general lack of classification of odontogenic tumors in veterinary medicine [30]. Histological similarities were observed with the previously reported odontogenic tumors in rats, wherein no ameloblastoma-like tumor cells were seen and ovoid neoplastic epithelial cells predominated [28]. But this was very different from DGCT in humans, wherein ameloblastomatous proliferation is typically obvious [3,17].

In megavertebrates, oral and facial proliferative lesions have been previously reported. This includes cases of gingivitis, tooth root abscessation, and SCC [31–33]. It is important to conduct routine clinical examinations and detailed histopathological examinations to properly diagnose these lesions. Despite its rarity, OGCL should be considered in cases of oral and facial proliferative lesions in megavertebrates and animals in general.

4. Conclusions

This is the first report of DGCT in veterinary medicine. The diagnosis of DGCT in this case was made based on routine histopathology, special histochemistry, and immunohistochemistry with human oncology and pathology as a reference. The histopathology and immunohistochemistry of DGCT in this rhinoceros match the majority of descriptions of DGCT in humans.

Author Contributions: Conceptualization, A.S. and M.Z.-S.; investigation, A.S., C.-R.J.; resources, A.S., Z.Z.Z.; data curation, Z.Z.Z., R.M.M.T., C.K.Y.; writing—original draft preparation, A.S.; writing—review and editing, M.Z-S., Z.Z.Z., C.-R.J. All authors have read and agreed to the published version of the manuscript.

Funding: This research was funded by the Federal Government of Malaysia, State Government of Sabah, and Sime Darby Foundation, and the APC was funded by Universiti Putra Malaysia.

Institutional Review Board Statement: Ethical review and approval were waived for this study because the study was non-experimental, and the samples analyzed were either part of clinical management or obtained in a necropsy.

Data Availability Statement: Not applicable.

Acknowledgments: The authors thank the staff of Borneo Rhino Alliance, especially the rhinoceros' keepers and field staff for attending to the rhinoceros at the Tabin Wildlife Reserve, as well as John Payne and Abdul Hamid Ahmad for their technical support. We also thank the veterinarians at the Sabah Wildlife Department for their assistances.

Conflicts of Interest: The authors declare no conflict of interest.

References

1. de Arruda, J.A.A.; Monteiro, J.L.G.C.; Abreu, L.G.; de Oliveira Silava, L.G.; Schuch, L.F.; de Noronha, M.S.; Callou, G.; Moreno, A.; Mesquita, R.A. Calcifying odontogenic cyst, dentinogenic ghost cell tumor, and ghost cell odontogenic carcinoma: A systematic review. *J. Oral Pathol. Med.* **2018**, *47*, 721–730. [CrossRef]
2. Flucke, U. Malignant Neoplasms of the Gnathic Bones. In *Head and Neck Pathology*, 3rd ed.; Thompson, L.D.R., Bishop, J.A., Eds.; Elsevier: Philadelphia, PA, USA, 2019; pp. 417–432.
3. Takata, T.; Slooweg, P.J. WHO classification of odontogenic and maxillofacial bone tumours. In *WHO Classification of Head and Neck Tumour*, 4th ed.; El-Naggar, A.K., Chan, J.K.C., Grandis, J.R., Takata, T., Slootweg, P.J., Eds.; International Agency for Research on Cancer: Lyon, France, 2017; pp. 204–260.
4. Bilodeau, E.A.; Seethala, R.R. Update on odontogenic tumors: Proceedings of the North American Head and Neck Pathology Society. *Head Neck Pathol.* **2019**, *13*, 457–465. [CrossRef]
5. Daniels, J.S.M. Recurrent calcifying odontogenic cyst involving the maxillary sinus. *Oral Surg. Oral Med. Oral Pathol. Oral Radiol. Endod.* **2004**, *98*, 660–664. [CrossRef] [PubMed]
6. Wright, B.A.; Bhardwaj, A.K.; Murphy, D. Recurrent calcifying odontogenic cyst. *Oral Surg. Oral Med. Oral Pathol. Oral Radiol. Endod.* **1984**, *58*, 579–583. [CrossRef]
7. Buchner, A.; Akrish, S.J.; Vered, M. Central dentinogenic ghost cell tumor: An update on a rare aggressive odontogenic tumor. *J. Oral Maxillofac. Surg.* **2016**, *74*, 307–314. [CrossRef]
8. Juneja, M.; George, J. Dentinogenic ghost cell tumor: A case report and review of the literature. *Oral Surg. Oral Med. Oral Pathol. Oral Radiol. Endod.* **2009**, *107*, e17–e22. [CrossRef] [PubMed]
9. Martos-Fernández, M.; Alberola-Ferranti, M.; Hueto-Madrid, J.A.; Bescós-Atín, C. Ghost cell odontogenic carcinoma: A rare case report and review of literature. *J. Clin. Exp. Dent.* **2014**, *6*, e602. [CrossRef]
10. Meuten, D.J.; Moore, F.M.; George, J.W. Mitotic count and the field of view area: Time to standardize. *Vet. Pathol.* **2016**, *53*, 7–9. [CrossRef] [PubMed]
11. Regezi, J.A. Odontogenic cysts, odontogenic tumors, fibroosseous, and giant cell lesions of the jaws. *Mod. Pathol.* **2002**, *15*, 331–341. [CrossRef]
12. Comolli, J.R.; Olsen, H.M.; Seguel, M.; Schnellbacher, R.W.; Fox, A.J.; Divers, S.J.; Sakamoto, K. Ameloblastoma in a wild black rat snake (*Pantherophis alleghaniensis*). *J. Vet. Diagn. Investig.* **2015**, *7*, 536–539. [CrossRef]
13. Del Corso, G.; Tardio, M.L.; Gissi, D.B.; Marchetti, C.; Montebugnoli, L.; Tarsitano, A. Ki-67 and p53 expression in ghost cell odontogenic carcinoma: A case report and literature review. *Oral Maxillofac. Surg.* **2015**, *19*, 85–89. [CrossRef] [PubMed]

14. Badger, K.V.; Gardner, D.G. The relationship of adamantinomatous craniopharyngioma to ghost cell ameloblastoma of the jaws: A histopathologic and immunohistochemical study. *J. Oral Maxillofac. Pathol.* **1997**, *26*, 349–355. [CrossRef] [PubMed]

15. Singhaniya, S.B.; Barpande, S.R.; Bhavthankar, J.D. Dentinogenic ghost cell tumor. *J. Oral Maxillofac. Pathol.* **2009**, *13*, 97. [CrossRef]

16. Ali, E.A.M.; Karrar, M.A.; El-Siddig, A.A.; Gafer, N.; Satir, A.A. Ghost cell odontogenic carcinoma of the maxilla: A case report with a literature review. *Pan Afr. Med. J.* **2015**, *21*. [CrossRef]

17. Lee, S.K.; Kim, Y.S. Current concepts and occurrence of epithelial odontogenic tumors: II. Calcifying epithelial odontogenic tumor versus ghost cell odontogenic tumors derived from calcifying odontogenic cyst. *Korean J. Pathol.* **2014**, *48*, 175. [CrossRef]

18. Folpe, A.L.; Tsue, T.; Rogersoin, L.; Weymuller, E.; Ods, D.; True, L.D. Odontogenic ghost cell carcinoma: A case report with immunohistochemical and ultrastructural characterization. *J. Oral Pathol. Med.* **1998**, *27*, 185–189. [CrossRef]

19. Piattelli, A.; Fioroni, M.; Di Alberti, L.; Rubini, C. Immunohistochemical analysis of a dentinogenic ghost cell tumour. *Oral Oncol.* **1998**, *34*, 502–507. [CrossRef]

20. Sekine, S.; Sato, S.; Takata, T.; Fukuda, Y.; Ishida, T.; Kishino, M.; Shibata, T.; Kanai, Y.; Hirohashi, S. β-catenin mutations are frequent in calcifying odontogenic cysts, but rare in ameloblastomas. *Am. J. Pathol.* **2003**, *163*, 1707–1712. [CrossRef]

21. Rappaport, M.J.; Showell, D.L.; Edenfield, W.J. Metastatic ghost cell odontogenic carcinoma: Description of a case and search for actionable targets. *Rare Tumors* **2015**, *7*, 96–97. [CrossRef] [PubMed]

22. Pinheiro, T.N.; de Souza, A.P.F.; Bacchi, C.E.; Consolaro, A. Dentinogenic ghost cell tumor: A bibliometric review of literature. *J. Oral Dis. Mark.* **2019**, *3*, 9–17. [CrossRef]

23. Kasai, T.; Kamegai, A.; Kubota, K.; Sato, K.; Kanematsu, N.; Mori, M. S-100 Protein Immunoreactivity of Calcifying/calcified Areas in Odontogenic Tumors. *Oral Med. Pathol.* **2002**, *7*, 19–25. [CrossRef]

24. Richardson, M.S.; Muller, S. Malignant odontogenic tumors: An update on selected tumors. *Head Neck Pathol.* **2014**, *8*, 411–420. [CrossRef]

25. Iezzi, G.; Rubini, C.; Fioroni, M.; Piattelli, A. Peripheral dentinogenic ghost cell tumor of the gingiva. *J. Periodontol.* **2007**, *78*, 1635–1638. [CrossRef] [PubMed]

26. Candido, G.A.; Viana, K.A.; Watanabe, S.; Vencio, E.F. Peripheral dentinogenic ghost cell tumor: A case report and review of the literature. *Oral Surg. Oral Med. Oral Pathol. Oral Radiol. Endod.* **2009**, *108*, e86–e90. [CrossRef]

27. Jayasooriya, P.R.; Mendis, B.R.R.N.; Lombardi, T. A Peripheral Dentinogenic Ghost Cell Tumor With Immunohistochemical Investigations and a Literature Review–Based Clinicopathological Comparison Between Peripheral and Central Variants. *Int. J. Surg. Pathol.* **2015**, *23*, 489–494. [CrossRef] [PubMed]

28. Cullen, J.M.; Ruebner, B.H.; Hsieh, D.P.H.; Burkers, E.J., Jr. Odontogenic tumors in Fischer rats. *J. Oral Pathol. Med.* **1987**, *16*, 469–473. [CrossRef]

29. Kang, M.S.; Park, M.S.; Kwon, S.W.; Ma, S.A.; Cho, D.Y.; Kim, D.Y.; Kim, Y. Amyloid-producing odontogenic tumour (calcifying epithelial odontogenic tumour) in the mandible of a Bengal tiger (*Panthera tigris tigris*). *J. Comp. Pathol.* **2006**, *134*, 236–240. [CrossRef] [PubMed]

30. Murphy, B.G.; Bell, C.M.; Soukup, J.W. Odontogenic Tumors. In *Veterinary Oral and Maxillofacial Pathology*; Wiley Blackwell: Hoboken, NJ, USA, 2020; pp. 91–128.

31. de Oliveira, A.R.; Arenales, A.; de Carvalho, T.P.; Pessanha, A.T.; Tinoco, H.P.; da Costa, M.E.L.T.; da Paixao, T.A.; Santos, R.L. Metastatic oral squamous cell carcinoma in a captive common hippopotamus (*Hippopotamus amphibius*). *Braz. J. Vet. Pathol.* **2018**, *11*, 64–67. [CrossRef]

32. Dennis, P.M.; Funk, J.A.; Rajala-Schultz, P.J.; Blumer, E.S.; Miller, R.E.; Wittum, T.E.; Saville, W.J. A review of some of the health issues of captive black rhinoceroses (*Diceros bicornis*). *J. Zoo Wildl. Med.* **2007**, *38*, 509–517. [CrossRef] [PubMed]

33. Langer, S.; Czerwonka, N.; Termes, K.; Herbst, W.; Koehler, K. Oral squamous cell carcinoma in an aged captive white rhinoceros (*Ceratotherium simum*). *J. Zoo Wildl. Med.* **2016**, *47*, 1090–1092. [CrossRef]

Case Report

Glial Response and Neuroinflammation in Cerebrocortical Atrophy in a Young Irish Wolfhound Dog

Fabiano J. F. de Sant'Ana [1,*], Miguel Omaña [2], Ester Blasco [3] and Martí Pumarola [3,4]

1   Laboratório de Diagnóstico Patológico Veterinário, Universidade de Brasília, Brasília 70636-020, Brazil
2   Hospital Veterinario Canis/MRIVETS, 07101 Palma de Mallorca, Spain; miguelomana@yahoo.es
3   Unit of Murine and Compared Pathology, Faculty of Veterinary, Universitat Autònoma de Barcelona, 08193 Bellaterra, Spain; ester.blasco@uab.cat (E.B.); marti.pumarola@uab.cat (M.P.)
4   Biomaterials and Nanomedicine (CIBER-BBN), Networking Research Center on Bioengineering, Universitat Autònoma de Barcelona, 08193 Bellaterra, Spain
*   Correspondence: santanafjf@yahoo.com; Tel.: +55-61-3468-7255

Simple Summary: Neuroinflammation is considered a reaction of the nervous system itself to protect and repair structural changes developed in it. Despite its initial positive purpose, sometimes it can produce worse consequences for the tissue. In this article, we present a case of an Irish Wolfhound dog suffering a rare idiopathic neurodegenerative disease producing wide cerebral cortical damage with loss of neuronal bodies in a bilateral and symmetrical pattern. We have studied the glial components of the neuroinflammation developed describing how they have exacerbated the nervous tissue damage.

Abstract: A two-year-old, Irish Wolfhound dog presented with a history of progressive neurological signs. Neurological exam revealed disorientation, absence of menace response, reduction of right nasal sensation, hypermetria and ataxia with reduction of proprioception in all four limbs. MRI findings were compatible with laminar neuronal necrosis and possible bilateral cortical cerebral atrophy. Grossly, a severe bilateral reduction of the gray matter with flattening of gyri, mainly in frontal and parietal cerebral areas, was observed. Histologically, multiple, segmental, bilateral, and symmetric areas of neuronal loss, necrosis and degeneration, in a laminar pattern, associated with a reactive gliosis were observed. Immunohistochemical studies showed severe reduction of neuronal bodies, proliferation and hypertrophy of astrocytes and microglia. Few perivascular B and T cells were demonstrated. Based on these data, we show some of the neuroinflammatory events that occur during CNS repair in a chronic phase of this condition.

Keywords: cerebral cortical atrophy; immunohistochemistry; neuronal necrosis; neuropathology

1. Introduction

Degenerative neurological diseases in domestic animals include a broad group of disorders that are characterized by progressive, bilateral and symmetrical degeneration and loss of cells, mainly neurons. Most of these diseases are of a genetic basis, but the precise pathogenic mechanisms are still poorly known or understood. Breed predisposition seems to occur in some cases. These diseases that affect the central nervous system can be classified in neuronal degenerations, axonal degenerations, myelin disorders, storage diseases, spongiform encephalopathies, spongy degenerations, and selective symmetrical encephalomalacias [1]. Recently, a novel idiopathic condition, characterized by superficial neocortical degeneration was reported in five dogs from North America and United Kingdom [2].

Neuroinflammation, the activation of the neuroimmune cells (microglia and astrocytes) into proinflammatory states, with no known causative insult and little change in blood-brain barrier biology, has been suggested as a pathological contributor in several

Citation: Sant'Ana, F.J.F.d.; Omaña, M.; Blasco, E.; Pumarola, M. Glial Response and Neuroinflammation in Cerebrocortical Atrophy in a Young Irish Wolfhound Dog. *Animals* **2021**, *11*, 143. https://doi.org/10.3390/ani11010143

Received: 6 October 2020
Accepted: 30 December 2020
Published: 11 January 2021

Publisher's Note: MDPI stays neutral with regard to jurisdictional claims in published maps and institutional affiliations.

neurodevelopmental, psychiatric, and neurodegenerative disorders [3]. This sustained inflammatory response suggests an important role of effectors of neuroinflammation in neuronal dysfunction and death [4]. Neuroinflammation is usually referred to as the chronic response of nervous tissue [3]. Neuroinflammation was previously understood as a local tissue response with few or no involvement of the peripheral immune system. Nevertheless, recent data support it is influenced by a number of peripheral and other factors, such as cytokines [5], chemokines expressed by lymphocytes [6], plasma kinins [7], hormones, and a complex of molecular interactions [4].

Here, we describe the pathological and immunohistochemical findings in one case of cerebrocortical atrophy in a young Irish Wolfhound dog, with emphasis on the glial reaction (neuroinflammation) and on the poor lymphocytic response to neuronal injury, which has not been previously investigated.

2. Description of the Case

A two-year-old, Irish Wolfhound, male dog presented with one-month history of progressive neurological signs, including difficulty to jump, lumbar pain and pelvic limb weakness. Hematological and biochemical (blood urea nitrogen, creatinine, glucose, albumin, total proteins, alanine aminotransferase, alkaline phosphatase) analyses did not reveal changes. Radiography of thorax was normal. In the neurological exam, disorientation, absence of menace response, reduction of right nasal sensation, hypermetria and ataxia with reduction of proprioceptive positioning in all four limbs with normal spinal reflexes were observed. These findings were indicative of diffuse primary bilateral cortico-thalamic lesions, with perhaps more severe involvement of the right cortical region, as well as the cerebellum. Magnetic resonance imaging (MRI) showed hyperintensity and increased width of the subarachnoid space surrounding the cerebral gyri in T2 weighted images. With similar distribution, these regions presented hypointensity in T1 and FLAIR (Fluid attenuated inversion recovery) sequences, and the cerebrospinal fluid sign was isointense. A line of hyperintensity was observed in the neocortical region in the FLAIR sequence, suggesting laminar neuronal necrosis (Figure 1). After the administration of contrast, there were zones of contrast enhancement in the neocortex and the meninges. MRI findings demonstrated possible bilateral cortical cerebral atrophy. The analysis of cerebrospinal fluid collected from the cerebellomedullary cistern revealed neutrophilic pleocytosis (50 cells/μL, 82% polymorphonucleates and 18% lymphocytes). RT-PCR analysis to six neurologic infectious diseases of dogs (canine distemper, toxoplasmosis, neosporosis, borreliosis, bartonellosis, and cryptococcosis) were negative. The animal was euthanized due to the poor prognosis and decision of the owner. Necropsy was performed and the brain was submitted to histopathological examination.

After fixation in 10% neutral buffered formalin for four days, transverse sections of the brain were performed. Representative fragments of cerebrum (frontal, parietal, temporal, occipital, and piriform cortical areas, basal nuclei, and hippocampus), thalamus, midbrain, pons, cerebellum and medulla oblongata were processed for routine histopathological examination upon hematoxylin and eosin staining. In addition, immunohistochemical (IHC) evaluation was performed using a biotin-peroxidase system and diaminobenzidine as the chromogen. Antigen retrieval was performed with citrate buffer pH 6.0 (NeuN, GFAP, Iba1 and CD20) or 0.1% protease (CD3). To block the endogenous peroxidase activity, the slides were incubated in a solution of H_2O_2 (3%) in distilled water. The reagents were applied manually, with an over-night incubation at 4 °C for the monoclonal primary antibodies and a 40 (NeuN, GFAP, CD3 and CD20) to 60 min (Iba1) incubation for the secondary antibodies. An avidin-biotin complex solution was used in the case of Iba1 to amplify the response and was incubated for 1 h. The diaminobenzidine chromogen was applied for 10 min. The IHC antibody panel is described in Table 1. The IHC sections were counterstained using Harris hematoxylin. The positive controls for IHC consisted of brain (NeuN, GFAP, and Iba-1) and lymph node (CD3 and CD20) of a dog without morphologic changes. For the negative controls, an isotype-specific immunoglobulin was used as a substitute for the primary antibody and no immunostaining was detected in these sections.

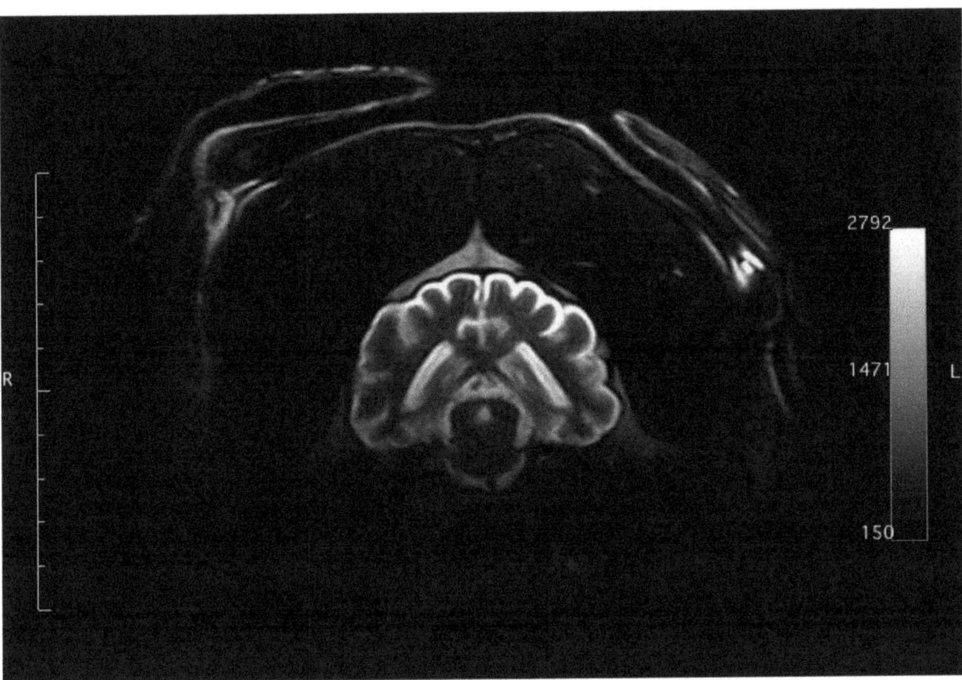

Figure 1. Brain magnetic resonance imaging from a young Irish Wolfhound dog with bilateral cerebrocortical atrophy. There is a line of hyperintensity in the neocortical region in the FLAIR transverse sequence, compatible with laminar neuronal necrosis.

Table 1. Immunohistochemical panel of antibodies used in this study.

Antibody	Reference	Manufacter	Dilution
NeuN (Clone A60)	MAB 377	Merck, Germany	1:100
Iba-1	ab5076	Abcan, Cambridge, UK	1:300
GFAP	Z0334	Dako, Denmark	1:5000
CD3	A0452	Dako, Denmark	1:100
CD20 (Clone MS4A1)	pa5-32313	Thermo Fisher Scientific, USA	1:300

Grossly, a severe reduction in neocortical gray matter with flattening of gyri bilaterally, affecting mainly frontal and parietal areas, was observed (Figure 2). In the surface of transverse sections, thinning of the frontal, parietal, and temporal cortices was more evident, and there was no clear distinction between gray and white matter. Gross changes in the occipital cortex, cerebellum, and brainstem were not observed.

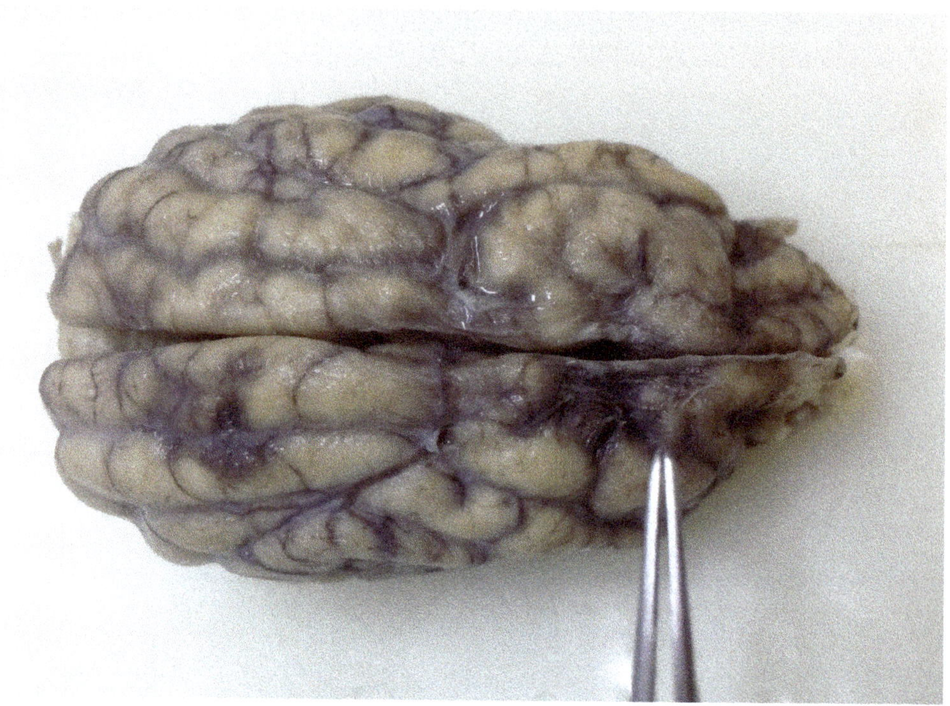

Figure 2. Dorsal view of the brain from a young Irish Wolfhound dog with bilateral cerebrocortical atrophy. There is atrophy and irregularity of gyri and enlargement of sulci.

Histologically, irregular, multiple, segmental areas of thinning of the gray matter with a bilateral and symmetrical pattern were observed in the frontal, parietal, temporal, and occipital cortices. Superficial layers (laminae I and II) were pallid due to severe absence of neuronal bodies and microspongiosis of neuropile (Figure 3A). In addition, individual neuronal necrosis and deposition of proteinaceous eosinophilic globules were observed. These changes were observed equally in the surface of the gyri and in deepest part of sulci. Numerous foamy macrophages (gitter cells) and a mild infiltration of lymphocytes was also noted, mainly in the subarachnoid space that was distended secondarily to cortical atrophy. In deepest layers (laminae IV and V), moderate depletion of neuronal bodies was associated with a reactive gliosis. Subcortical white matter showed disorganization and mild to moderate spongiosis. In the hippocampus, individual neuronal necrosis with reactive gliosis was observed in the CA2 and CA4 regions. Significant lesions were not detected in the other regions evaluated. The severe decrease in number of neuronal bodies in the affected cortex was confirmed by NeuN immunostaining. The few layers with remaining neurons were disorganized (Figure 3B). Glial Fibrillary Acidic Protein (GFAP) labelling showed evident proliferation and hypertrophy of astrocytes in the affected gray matter, mainly in the more superficial (adjacent to pia mater) and also in the deepest layers. Reactive astrocytes frequently presented enlarged nuclei and abundant and extended cytoplasmic processes (Figure 3C). Proliferation and hypertrophy of immunopositive Iba-1 microglia cells were also evident in the affected gray matter areas (Figure 3D). Hypertrophied microglia were observed mainly in the deeper neuronal layers and in the subarachnoid space. CD20 and CD3 immunolabeling demonstrated the presence of few perivascular B and T cells, respectively, located in the subarachnoid space and rarely in the subpial areas of the affected cortex.

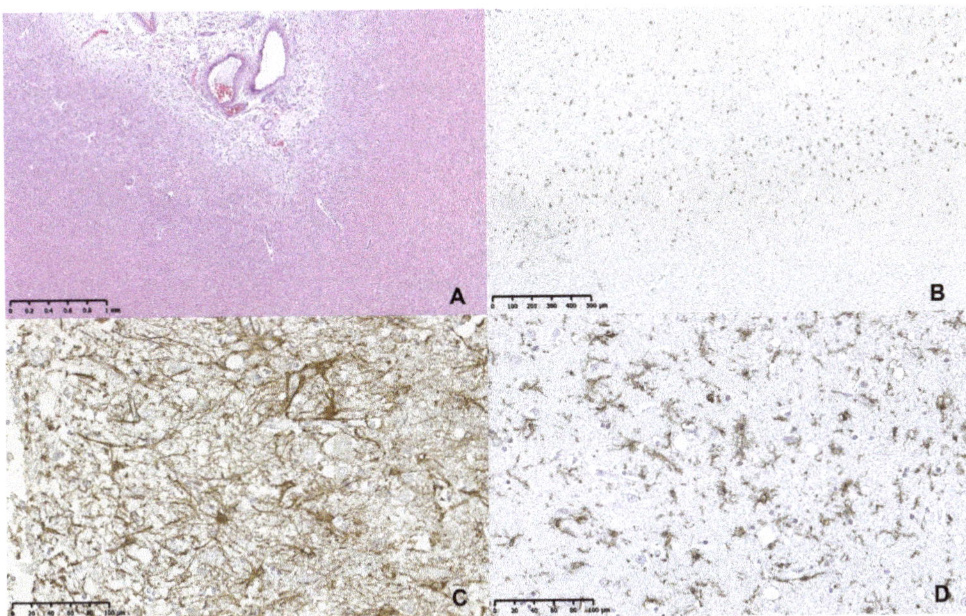

Figure 3. Brain histopathology and immunohistochemistry from a young Irish Wolfhound dog with bilateral cerebrocortical atrophy. (**A**). There is distension and diffuse cell infiltration of the subarachnoid space and superficial areas of pallor affecting upper cortical laminae with loss of the grey matter. Parietal cortex, HE. (**B**). Note the loss of neuronal bodies and cortical architecture disorganization. NeuN immunohistochemistry. Hematoxylin counterstain. (**C**). Numerous reactive and hypertrophied astrocytes are noted in the affected neocortical areas. GFAP immunohistochemistry. Hematoxylin counterstain. (**D**). There are abundant reactive microglia cells in the injured cortex. Iba-1 immunohistochemistry. Hematoxylin counterstain.

3. Discussion

The current manuscript describes the neurohistological and immunohistochemical study of an unusual case of cerebrocortical chronic selective bilateral laminar neuronal degeneration and necrosis in a two-year-old dog. A recent retrospective study (1982–2012) described a very similar neuropathological condition in five young dogs of both sexes from United Kingdom and USA [2]. Three out of these five dogs were related hounds (two Irish Wolfhound and one Scottish Deerhound). In the present study, the affected dog was an Irish Wolfhound that lived in Spain, though it was initially acquired in Germany. These data indicate a strong probability of a hereditary and genetic basis for this disease, although the pathogenesis of the condition of insidious onset remains unknown. Few inherited disorders have been described and recognized in Irish Wolfhound dogs, such as dilated cardiomyopathy, hip dysplasia, and portosystemic shunts [8–11]. In addition, hyperekplexia (Startle disease), a rare severe congenital disorder, has been recognized as one of the few inherited neurological diseases diagnosed in this canine breed, usually in young puppies [7].

Clinical signs of this chronic neurodegenerative disorder of young dogs are progressive and include mainly ataxia, paresis, hyperextension of limbs, blindness, difficulty of prehension, seizures, and hyperesthesia presented over weeks to months [2]. Some of these signs were observed in the current case. Our MRI findings are in accordance to the ones reported previously [2]. Based on the imaging findings, we could not objectively determine cortical atrophy due to the limited number of studies on cortical thickness in dogs. Clinical history and neurological exam were suggestive of diffuse cerebral lesion affecting the neocortex (sensory and visual deficits) and brainstem (consciousness or proprioceptive deficits), though the MRI only suggested a severe neocortical degenerative and/or necrotic lesion. Although hypermetria was suggestive of cerebellar involvement, we could not find any structural change in this region to explain this clinical sign. The cerebellum is a structure commonly affected in neurodegenerative diseases of domestic animals, mainly in cerebellar cortical abiotrophies (1). There were no scientific evidences to explain the high selectivity of some neuronal groups in these diseases. Systemic infectious diseases or metastasis affecting the CNS can cause multifocal lesions and similar clinical signs. However, the negative results of PCR investigations performed in this study for six important systemic diseases of dogs and the image findings made less probable these potential causes.

The histopathological pattern of this condition is considered atypical, because usually in domestic animals with cerebrocortical degeneration and necrosis, several layers of neurons are injured [12,13]. Here, we observed a chronic and selective neuronal loss mainly affecting superficial (and eventually deeper) layers of the neocortex and hippocampus, as previously reported [2]. Occasionally, other regions of the brain, such as hypothalamus, mesencephalon, and pons can also be affected [2], but no changes were detected in our case. Our histopathological findings indicate an expressive response of astrocytes and microglial cells to the chronic severe neuronal damage. The proliferation and activation of both glial cells in this neurodegenerative condition were considered crucial to the repair, regulation, phagocytosis and structural support of the injured tissue. These cells provide pre- and anti-inflammatory actions in various functions under basal and disease conditions (3).

In the current study, immunohistochemical evaluation was effective to demonstrate the exclusive glial response to neuronal changes, characterized by proliferation and hypertrophy of astrocytes and microglia, particularly in the most affected areas of neocortex. These findings are associated with the chronic progressive loss and injury of cortical neurons highlighted by labelling with NeuN. In the current study, the lymphocyte-poor immune response to the neuronal damage was restricted to subarachnoid space and subpial areas. The findings of the current study suggest that an initial glial response produced to control and repair the neuronal damage was converted in an exaggerated response (neuroinflammation) and negative to the nervous system, with increased tissue injury. Future studies evaluating the complex neuroinflammatory pathways and genetic features of this uncommon neurodegenerative disorder are needed to further characterize its etiology and pathogenesis. The fact that inflammation can be either protective or damaging, is fundamental to understand the pathogenesis of brain disorders.

4. Conclusions

Based on histopathological and immunohistochemical results of the current study, we suggest that exclusive glial response, including proliferation and hypertrophy of astrocytes and microglia, is a crucial event in the chronic neuronal damage, observed in the selective superficial neocortical neuronal degeneration of Irish Wolfhound dogs, without effective involvement of the lymphocytic response.

Author Contributions: All authors have contributed equally, read and agreed to the final version of the manuscript. All authors have read and agreed to the published version of the manuscript.

Funding: This research received no external funding.

Institutional Review Board Statement: Ethical review and approval were waived for this study, because the samples analyzed were obtained in a necropsy of a dog that was euthanized after decision of the owner.

Informed Consent Statement: Not applicable.

Data Availability Statement: Not applicable.

Acknowledgments: The authors thank Tamara Rivero Balsera for the technical assistance and Ronaldo C. da Costa (OSU) for the critical review of the manuscript.

Conflicts of Interest: The authors declare no conflict of interest.

References

1. Vandevelde, M.; Higgins, R.J.; Oevermann, A. *Veterinary Neuropathology, Essentials of Theory and Practice*; Wiley: Chichester, UK, 2012; pp. 157–192.
2. Cahalan, S.D.; Cappello, R.; de Lahunta, A.; Summers, B.A. A novel idiopathic superficial neocortical degeneration and atrophy in young adult dogs. *Vet. Pathol.* **2015**, *52*, 344–350. [CrossRef] [PubMed]
3. Schain, M.; Kreisl, W.C. Neuroinflammation in neurodegenerative disorders—A review. *Curr. Neurol. Neurosci. Rep.* **2017**, *17*, 25. [CrossRef] [PubMed]
4. Niranjan, R. Recent advances in the mechanisms of neuroinflammation and their roles in neurodegeneration. *Neurochem. Int.* **2018**, *120*, 13–20. [CrossRef] [PubMed]
5. Adalid-Peralta, L.; Fleury, A.; García-Ibarra, T.M.; Hernández, M.; Parkhouse, M.; Crispín, J.C.; Voltaire-Proaño, J.; Cárdenas, G.; Fragoso, G.; Sciutto, E. Human neurocysticercosis: In vivo expansion of peripheral regulatory T cells and their recruitment in the central nervous system. *J. Parasitol.* **2012**, *98*, 142–148. [CrossRef] [PubMed]
6. Adzemovic, M.Z.; Öckinger, J.; Zeitelhofer, M.; Hochmeister, S.; Beyeen, A.D.; Paulson, A.; Gillett, A.; Hedreul, M.T.; Covacu, R.; Lassmann, H.; et al. Expression of Ccl11 associates with immune response modulation and protection against neuroinflammation in rats. *PLoS ONE* **2012**, *7*, e39794. [CrossRef] [PubMed]
7. Gill, J.L.; Capper, D.; Vanbellinghen, J.; Chung, S.-K.; Higgins, R.J.; Rees, M.I.; Shelton, G.D.; Harvey, R.J. Startle disease in Irish wolfhounds associated with a microdeletion in the glycine transporter GlyT2 gene. *Neurobiol. Dis.* **2011**, *43*, 184–189. [CrossRef] [PubMed]
8. Canine Inherited Disorders Database (CIDD) Irish Wolfhound. University of Prince Edward Island, 2018. Available online: https://cidd.discoveryspace.ca/breed/irish-wolfhound.html (accessed on 19 December 2019).
9. Krontveit, R.I.; Trangerud, C.; NØdtvedt, A.; Dohoo, I.; Moe, L.; Saevik, B.K. The effect of radiological hip dysplasia and breed on survival in a prospective cohort study of four large dog breeds followed over a 10 year period. *Vet. J.* **2012**, *193*, 206–211. [CrossRef] [PubMed]
10. Spier, A.W.; Meurs, K.M.; Coovert, D.D.; Lehmkuhl, L.B.; O'Grady, M.R.; Freeman, L.M.; Burghes, A.H.; Towbin, J.A. Use of western immunoblot for evaluation of myocardial dystrophin, alpha-sarcoglycan, and beta-dystroglycan in dogs with idiopathic dilated cardiomyopathy. *Am. J. Vet. Res.* **2001**, *62*, 67–71. [CrossRef] [PubMed]
11. Zandvliet, M.M.; Rothuizen, J. Transient hyperammonemia due to urea cycle enzyme deficiency in Irish Wolfhounds. *J. Vet. Intern. Med.* **2007**, *21*, 215–218. [CrossRef] [PubMed]
12. Miller, A.D.; Zachary, J.F. Nervous system. In *Pathologic Basis of Veterinary Disease*, 6th ed.; Zachary, J.F., Ed.; Elsevier: St. Louis, MO, USA, 2017; pp. 805–907.
13. Palmer, A.C.; Walker, R.G. The neuropathological effects of cardiac arrest in animals: A study of five cases. *J. Small Anim. Pract.* **1970**, *11*, 779–791. [CrossRef] [PubMed]

Article

The Influence of *Fusarium* Mycotoxins on the Liver of Gilts and Their Suckling Piglets

Tamara Dolenšek [1],*, Tanja Švara [1], Tanja Knific [2], Mitja Gombač [1], Boštjan Luzar [3] and Breda Jakovac-Strajn [2]

[1] Institute of Pathology, Wild Animals, Fish and Bees, Veterinary Faculty, University of Ljubljana, Gerbičeva ulica 60, 1000 Ljubljana, Slovenia; tanja.svara@vf.uni-lj.si (T.Š.); mitja.gombac@vf.uni-lj.si (M.G.)

[2] Institute of Food Safety, Feed and Environment, Veterinary Faculty, University of Ljubljana, Gerbičeva ulica 60, 1000 Ljubljana, Slovenia; tanja.knific@vf.uni-lj.si (T.K.); breda.jakovac-strajn@vf.uni-lj.si (B.J.-S.)

[3] Institute of Pathology, Faculty of Medicine, University of Ljubljana, Korytkova 2, 1000 Ljubljana, Slovenia; bostjan.luzar@mf.uni-lj.si

* Correspondence: tamara.dolensek@vf.uni-lj.si

Simple Summary: Mycotoxins are toxic secondary metabolites of fungi that frequently contaminate animal feed and human food in different combinations; therefore, it is of great importance to determine the effects of mycotoxin co-contamination. Pigs are one of the most sensitive animal species to *Fusarium* mycotoxins, and the liver is an important site of mycotoxin metabolism. The objective of the present research was to determine histopathological changes, apoptosis, and proliferation in the liver of gilts fed with *Fusarium* mycotoxin-contaminated feed for a prolonged time at the end of their pregnancy and until weaning of their piglets. Additionally, the same parameters were evaluated in the liver of their piglets to determine whether *Fusarium* mycotoxins would affect the offspring. The results revealed increased hepatocellular necrosis and apoptosis as well as sinusoidal leukocytosis with inflammatory infiltrates of hepatic lobules in experimental gilts, but no significant changes were observed in the piglet livers, implying that the utilized concentrations and duration of exposure did not cause detrimental effects on them. Interestingly, the amount of interlobular connective tissue in the liver of experimental gilts was significantly decreased. The obtained results emphasized the need to evaluate *Fusarium* mycotoxin concentrations in feed because even at low concentrations, they can cause adverse effects, but there is less concern for severe detrimental effects on the offspring.

Citation: Dolenšek, T.; Švara, T.; Knific, T.; Gombač, M.; Luzar, B.; Jakovac-Strajn, B. The Influence of *Fusarium* Mycotoxins on the Liver of Gilts and Their Suckling Piglets. *Animals* **2021**, *11*, 2534. https://doi.org/10.3390/ani11092534

Academic Editor: Alejandro Suarez Bonnet

Received: 14 May 2021
Accepted: 26 August 2021
Published: 28 August 2021

Abstract: Mycotoxins are common fungal secondary metabolites in both animal feed and human food, representing widespread toxic contaminants that cause various adverse effects. Co-contamination with different mycotoxins is frequent; therefore, this study focused on feed contaminated with *Fusarium* mycotoxins, namely, deoxynivalenol (5.08 mg/kg), zearalenone (0.09 mg/kg), and fusaric acid (21.6 mg/kg). Their effects on the liver of gilts and their piglets were chosen as the research subject as pigs are one of the most sensitive animal species that are also physiologically very similar to humans. The gilts were fed the experimental diet for 54 ± 1 day, starting late in their pregnancy and continuing until roughly a week after weaning of their piglets. Livers of gilts and their piglets were assessed for different histopathological changes, apoptosis, and proliferation activity of hepatocytes. On histopathology, gilts fed the experimental diet had a statistically significant increase in hepatocellular necrosis and apoptosis ($p = 0.0318$) as well as sinusoidal leukocytosis with inflammatory infiltrates of hepatic lobules ($p = 0.0004$). The amount of interlobular connective tissue in the liver of experimental gilts was also significantly decreased ($p = 0.0232$), implying a disruption in the formation of fibrous connective tissue. Apoptosis of hepatocytes and of cells in hepatic sinusoids, further assessed by the terminal deoxynucleotidyl transferase dUTP nick-end labelling (TUNEL) assay, showed a statistically significant increase ($p = 0.0224$ and $p = 0.0007$, respectively). No differences were observed in piglet livers. These results indicated that *Fusarium* mycotoxins elicited increased apoptosis, necrosis, and inflammation in the liver of gilts, but caused no effects on the liver of piglets at these concentrations.

Keywords: *Fusarium* mycotoxins; pig; liver; histopathology; immunohistochemistry; apoptosis; proliferation index

1. Introduction

Mycotoxins are toxic secondary metabolites produced by many filamentous fungi. The most important fungi causing frequent and problematic contamination of human food and animal feed belong to the fungal genera of *Fusarium*, *Aspergillus*, and *Penicillium*. Maize is considered the most susceptible crop for mycotoxin contamination and rice the least susceptible one [1].

Fusarium fungi produce a variety of toxic secondary metabolites, which are not essential to fungal growth but can induce several adverse effects in livestock [2]. The most toxicologically important *Fusarium* toxins are fumonisins (FBs), zearalenone (ZEN), and trichothecenes, such as deoxynivalenol (DON), nivalenol (NIV), diacetoxyscirpenol (DAS), and T-2 toxin [3]. Both in vitro and in vivo studies have demonstrated that toxicokinetics, bioavailability, and the mechanisms of action of these substances vary depending on the species involved [4]. ZEN causes reproductive abnormalities in pigs and ruminants and DON is well known for being a potent feed intake inhibitor in pigs [5,6]. Next to these well-known *Fusarium* mycotoxins, there are also several unregulated, so-called emerging mycotoxins, which frequently occur in agricultural products. One of them is fusaric acid (FA), which is found in several types of cereal grain and mixed feeds. This mycotoxin needs to be further investigated in vitro and in vivo because its neurochemical effects and possible synergistic effects with other mycotoxins, especially DON and FBs, may pose a problem to humans and livestock [7].

Besides aflatoxins (AFs) and ochratoxins, which are not *Fusarium* mycotoxins, FBs, ZEN, and trichothecenes, especially DON, are considered highly important in food safety and public health due to their widespread occurrence and toxicity. In people, chronic exposure to mycotoxins, even at low levels, may lead to adverse effects in different organs, such as the liver, kidneys, and immune system [8,9].

Due to the frequent presence of several different mycotoxins in grain and animal feed, widespread reports of co-contamination are of great potential significance [10]. A global survey indicated that 72% of samples of feed and feed raw materials are positive for at least one mycotoxin and 38% are co-contaminated [11], whereas several studies in European countries, simultaneously analyzing 20 or more mycotoxins, have shown a remarkable 44–100% of such samples to be co-contaminated with more than one mycotoxin [12]. A recent study that included 524 worldwide finished pig feed samples detected more than 235 different metabolites, including regulated mycotoxins, emerging mycotoxins, and modified/masked mycotoxins. DON was detected in 88% of the samples, mostly from the Northern Hemisphere, with a median concentration of 0.206 mg/kg of feed. All DON-contaminated samples were co-contaminated by other mycotoxins, the second most common being ZEN with a median concentration of 0.018 mg/kg of feed, while FA was not among the 60 most prevalent fungal metabolites [13].

Concomitantly occurring mycotoxins can have antagonistic, additive, or synergistic effects [14], but very little is known about their potential interactive toxic effects [2]. Even though the results from the global survey indicated that the *Fusarium* mycotoxins DON, FBs, and ZEN contaminated 55%, 54%, and 36% of feed and feed ingredients, respectively, most samples complied with even the most rigorous European Union regulations or recommendations on the maximal tolerable concentrations of individual mycotoxins [11]. Currently, the European Commission's recommendation and its amendment on the presence of DON, ZEN, ochratoxin A, T-2 and HT-2, and FBs in products intended for animal feeding suggest that compound feed for piglets and gilts does not exceed 0.9 mg of DON/kg, 0.1 mg of ZEN/kg, 0.05 mg of ochratoxin A/kg and 5 mg of fumonisins B1 + B2/kg [15,16]. It is therefore of great importance to determine the effects of co-contaminating *Fusarium* mycotoxins, especially at naturally occurring concentrations, as well as concentrations lower than the accepted tolerance concentrations for individual mycotoxins.

Pigs are especially interesting for further research because they are one of the most sensitive animal species for *Fusarium* mycotoxins, especially ZEN and trichothecenes, such as DON and T-2. They are usually fed a cereal-rich diet, which can expose them to higher levels of these mycotoxins. As they are physiologically very similar to humans, they can serve as a good translational animal model, especially due to their similarities in the intestinal tract [17]. The effects of these toxins partly depend on their absorption, distribution, metabolism, and excretion (ADME processes) by the animal species in question. As the ADME processes seem to be qualitatively quite similar between pigs and humans, pigs can be very useful in the risk assessment of mycotoxins and for establishing legal limits of mycotoxins [18].

Research on the effects of feeding pigs with *Fusarium* mycotoxin co-contaminated feed has been ongoing for over 30 years, providing insight in various aspects. These studies often emphasized the zootechnical, hematological, biochemical, toxicological, and immunological parameters [19–23], whereas others also investigated histological changes in various organs with or without the aid of immunohistochemistry [14,24–35] or even examined gene expression profiles [36–38].

Since *Fusarium* mycotoxins are such a common contaminant and clearly have effects on different animal species, it is also of interest whether they have detrimental effects on the offspring. Some studies have analyzed the transfer of single or multiple *Fusarium* mycotoxins from sows to their offspring, implying that these can cause indirect effects via a decreased feed intake and via direct effects of diaplacentar transfer of ingested mycotoxins to the developing fetuses [20,21,29–31,39,40].

The liver is an important site of *Fusarium* mycotoxin metabolism [18]. DON's effects on liver have been investigated by studies evaluating biochemical, functional, histopathological parameters [33,34,41–43] and even gene expression profiles [38].

The aim of this study was therefore to determine whether feed containing naturally occurring concentrations of DON, ZEN, and FA would elicit histopathological changes, a difference in the number of apoptotic cells, and the proliferation index in the liver of gilts and their suckling piglets.

2. Materials and Methods

2.1. Research Design

This study was conducted on samples retrieved from the experiment approved by the Veterinary Administration of the Republic of Slovenia and described in detail by Jakovac-Strajn et al. [22]. In summary, the experiment included 10 gilts that were fed an experimental diet containing maize naturally contaminated with *Fusarium* mycotoxins, 10 gilts that were fed a control diet, and the offspring of both groups. The gilts were daily fed 3.5 kg of the diet during gestation and 6 kg of the same diet from the day of farrowing until weaning. The gilts from the experimental group consumed significantly less than the control group, but the average bodyweight was not significantly different even at the end of the experiment. At the start of the experiment, the gilts were at 89 ± 2 days of gestation, and they remained in the experiment for a total of 54 ± 1 day. The farrowing in both groups started 24 to 27 days after the start of the experiment and the piglets were weaned at 21 days of age. No antimicrobials were given to either the gilts or their piglets during the experiment.

The experimental diet contained 5.08 mg DON, 0.09 mg ZEN, and 21.6 mg FA per kg of feed. The control diet contained 0.29 mg DON per kg of feed, whereas ZEN (<0.02 mg/kg) and FA (<0.77 mg/kg) were below their detection limits. The concentrations of aflatoxin B1 (<0.2 µg/kg), 15-ADON, NIV, fusarenon-X, DAS, T-2 toxin, HT-2 toxin (<0.05 µg/kg), ochratoxin A, and fumonisins B1, B2, and B3 (<10 µg/kg) were also measured in both diets, but they were all below their detection limits, these being indicated in parentheses.

In order to collect organs for further examination, a single 7-day-old suckling piglet was randomly selected from each of the 20 litters and killed by lethal injection of T-61 solution (embutramide/mebezonium iodide/tetracaine hydrochloride, Intervet, Unterschleißheim, Germany), whereas all the gilts were killed by captive bolt and exsanguination 5 to 8 days after weaning of the remaining piglets in the litters. Afterwards, liver samples were immediately collected, fixed in 10% phosphate buffered formalin and routinely embedded in paraffin blocks. Liver samples from two killed suckling piglets, one from each group, were inappropriate for further processing.

2.2. Histopathology of the Liver of Gilts and Their Suckling Piglets

Histopathological examination of 4 μm thick tissue sections of formalin-fixed paraffin-embedded (FFPE) liver samples stained with hematoxylin and eosin (H&E) was conducted using light microscopy. Several different histopathological changes were assessed in the liver: irregularity of hepatic cords, fibrosis, sinusoidal leukocytosis with inflammatory infiltrates of hepatic lobules, portal tract inflammatory infiltrates, hepatocytes with vacuolar or granular cytoplasm, hepatocellular necrosis and apoptosis, markedly enlarged hepatocytes (hepatocellular megalocytosis), markedly enlarged hepatocellular nuclei (hepatocellular megakaryosis), biliary hyperplasia, dilatation and thickening of blood vessels, and thrombosis of blood or lymphatic vessels.

Each assessed histopathological change was graded for its intensity and extent. The intensity of the histopathological changes was assigned one of the following scores: 0—not present, 1—mild, 2—moderate, and 3—severe. The extent of the histopathological changes was assigned one of the following scores: 0—not present, 1—minimal (0 to <5% of the tissue section), 2—mild (5 to <15% of the tissue section), 3—moderate (15 to <40% of the tissue section) and 4—severe (40% or more of the tissue section). The assigned intensity and extent score were then multiplied to obtain the final score for each histopathological change in the tissue section of each liver sample from both the gilts and their piglets.

2.3. Detection of Apoptotic Cells in the Liver of Gilts and Their Suckling Piglets

For the detection of apoptotic cells in 4 μm thick FFPE tissue sections of liver from both the gilts and their piglets, we performed the terminal deoxynucleotidyl transferase dUTP nick-end labelling (TUNEL) assay using a commercial kit (ApopTag® peroxidase in situ apoptosis detection kit; Chemicon, Temecula, CA, USA) according to the manufacturer's instructions. Finally, the tissue sections were counterstained with Mayer's hematoxylin and coverslipped. Tissue sections of porcine kidney incubated with RQ1 RNase-Free DNase (M6101; Promega, Madison, WI, USA) were used as the positive control, and tissue sections of porcine kidney that were only incubated with the label solution (without terminal deoxynucleotidyl transferase) served as the negative control.

Using light microscopy, we counted TUNEL-positive cells in 30 randomly selected high-power fields (HPF), and also noted whether they were apoptotic hepatocytes or apoptotic cells in hepatic sinusoids. For hepatocytes to be considered TUNEL-positive, they had to have a clearly stained nucleus. Similarly, apoptotic cells in hepatic sinusoids had to exhibit moderate to marked nuclear staining.

2.4. Determining the Proliferation Index in the Liver of Gilts and Their Suckling Piglets

The proliferation activity of hepatocytes was evaluated on 4 μm thick FFPE tissue sections of liver from both the gilts and their piglets using immunohistochemical labelling with the mouse monoclonal antibody raised against human Ki-67 antigen, clone MIB-1 (Dako, Glostrup, Denmark), which was diluted 1:75. The antigen retrieval was performed by microwave treatment at a medium power (550 W) for 15 min in ethylenediaminetetraacetic acid (EDTA) with a pH of 8.0. The tissue sections were then incubated with primary antibodies for 1 hour at room temperature in a humid chamber. Endogenous peroxidase activity was quenched in the peroxidase-blocking solution Dako REALTM (Dako, Glostrup, Denmark) for 30 min at room temperature. The visualization kit Dako REALTM EnVision™ Detection System Peroxidase/DAB+, Rabbit/Mouse (Dako, Glostrup, Denmark) was applied according to the manufacturer's instructions. Finally, the tissue sections were counterstained with Mayer's hematoxylin and coverslipped. Tissue sections of porcine spleen were used as the positive control, and tissue sections of porcine liver that were not treated with primary antibodies served as the negative control. From one of the experimental gilts, a tissue section of the liver was not acquired due to lack of adequate FFPE tissue.

The proliferation index of hepatocytes was calculated as the rate of Ki-67-positive nuclei in a total of 1000 counted nuclei in the tissue sections of liver under a light microscope.

2.5. Morphometrical Evaluation of Interlobular Connective Tissue in the Liver of Gilts

The amount of interlobular connective tissue was measured in 4 μm thick FFPE tissue sections of liver samples only from the gilts. The tissue sections were stained with Goldner's Masson trichrome stain to clearly depict fibrous connective tissues under a light microscope coupled with a digital camera. Using the software program NIS-Elements Basic Research (Nikon Instruments Inc., Tokyo, Japan), five consecutive microphotographs at HPF were made for each tissue section and represented the area of measurement. The microphotographs were then converted into a binary-colored output by marking pixels that belonged to either interlobular connective tissue or parenchyma, thus enabling automated detection of interlobular connective tissue. The amount of interlobular connective tissue was expressed as the area fraction of the corresponding pixels out of the total number of pixels in the area of measurement. When necessary, the automatically detected areas of interlobular connective tissue were corrected manually.

2.6. Statistical Analysis

For statistical analysis, we used the R statistical software, version 3.6.2 (R Foundation for Statistical Computing, Vienna, Austria) [44]. The obtained results for the experimental and control groups of both the gilts and their piglets are presented with basic descriptive statistics. The Shapiro–Wilk test was used to assess the normality of the variables. For both gilts and piglets, the differences between the experimental and control group were analyzed with the two-tailed Mann–Whitney U test because most of the variables had a non-normal distribution. The correlations between histopathological changes, the number of apoptotic cells and the proliferation index were assessed separately for the gilts and their piglets with Spearman's rank correlation coefficients and Holm's adjusted p-values. Statistical significance was determined as $p < 0.05$, and $0.05 \leq p < 0.1$ was marginally significant.

3. Results

3.1. Histopathology of the Liver of Gilts and Their Suckling Piglets

In gilts, seven assessed histopathological changes were observed on H&E-stained liver sections: fibrosis, sinusoidal leukocytosis with inflammatory infiltrates of hepatic lobules, portal tract inflammatory infiltrates, hepatocytes with vacuolar or granular cytoplasm, hepatocellular necrosis and apoptosis, hepatocellular megalocytosis, and hepatocellular megakaryosis. Overt fibrosis was only observed in the liver of one gilt from the control group, but even this was mild based on its final score of 1 and was not statistically significant in comparison with the experimental group ($p = 0.3681$). The remaining six histopathological changes also had low final scores with a maximum score of 2, and in one experimental gilt, hepatocytes with vacuolar or granular cytoplasm reached a score of 4 (Table 1). Final scores for hepatocellular necrosis and apoptosis ($p = 0.0318$) and sinusoidal leukocytosis with inflammatory infiltrates of hepatic lobules ($p = 0.0004$) were significantly higher in the experimental group compared with the control group (Figure 1A,B), whereas portal tract inflammatory infiltrates ($p = 0.4539$) showed no statistically significant differences. Additionally, hepatocellular necrosis and apoptosis and sinusoidal leukocytosis with inflammatory infiltrates of hepatic lobules were strongly correlated (Spearman's $\rho = 0.73, p = 0.0226$). The inflammatory infiltrates were composed of different proportions of lymphocytes, plasma cells, neutrophils, eosinophils, and, occasionally, macrophages. Hepatocytes with vacuolar or granular cytoplasm ($p = 0.3681$), hepatocellular megalocytosis ($p = 1.0000$), and hepatocellular megakaryosis ($p = 1.0000$) did not show statistically significant differences between both groups.

Table 1. The final scores for each histopathological change in the liver of control and experimental gilts presented with basic descriptive statistics.

Histopathological Change	Group	N	Min	Q_1	Median	Q_3	Max
Fibrosis	control	10	0	0	0	0	1
	experimental	10	0	0	0	0	0
Sinusoidal leukocytosis with inflammatory infiltrates of hepatic lobules *	control	10	0	0	0	0	0
	experimental	10	0	1	1	1	1
Portal tract inflammatory infiltrates	control	10	0	1	1	1	2
	experimental	10	0	1	1	1	2
Hepatocytes with vacuolaror or granular cytoplasm	control	10	0	0	0	0	0
	experimental	10	0	0	0	0	4
Hepatocellular necrosis and apoptosis *	control	10	0	0	0	0.75	1
	experimental	10	0	1	1	1	1
Hepatocellular megalocytosis	control	10	0	0	0	0	1
	experimental	10	0	0	0	0	1
Hepatocellular megakaryosis	control	10	0	0	0	0	1
	experimental	10	0	0	0	0	1

* Statistically significant difference between the control and experimental groups ($p < 0.05$). N—number of animals, Min—minimum, Q_1—lower quartile, Q_3—upper quartile, Max—maximum.

In piglets, only three assessed histopathological changes were observed on H&E-stained liver sections, these being hepatocytes with vacuolar or granular cytoplasm, hepatocellular megalocytosis, and hepatocellular megakaryosis. Hepatocellular megalocytosis and hepatocellular megakaryosis had low final scores that did not exceed a final score of 1, whereas hepatocytes with vacuolar or granular cytoplasm had a final score ranging between 0 and 12 (Table 2). Similar to what was observed in gilts, hepatocytes with vacuolar or granular cytoplasm ($p = 0.1981$), hepatocellular megalocytosis ($p = 1.0000$), and hepatocellular megakaryosis ($p = 1.0000$) showed no statistically significant differences between both groups (Figure 1C,D).

Table 2. The final scores for each histopathological change in the liver of control and experimental piglets presented with basic descriptive statistics.

Histopathological Change	Group	N	Min	Q_1	Median	Q_3	Max
Hepatocytes with vacuolar or granular cytoplasm	control	9	0	4	8	8	12
	experimental	9	0	0	4	4	8
Hepatocellular megakaryosis	control	9	0	0	0	0	1
	experimental	9	0	0	0	0	1
Hepatocellular megakaryosis	control	9	0	0	0	1	1
	experimental	9	0	0	0	1	1

N—number of animals, Min—minimum, Q_1—lower quartile, Q_3—upper quartile, Max—maximum.

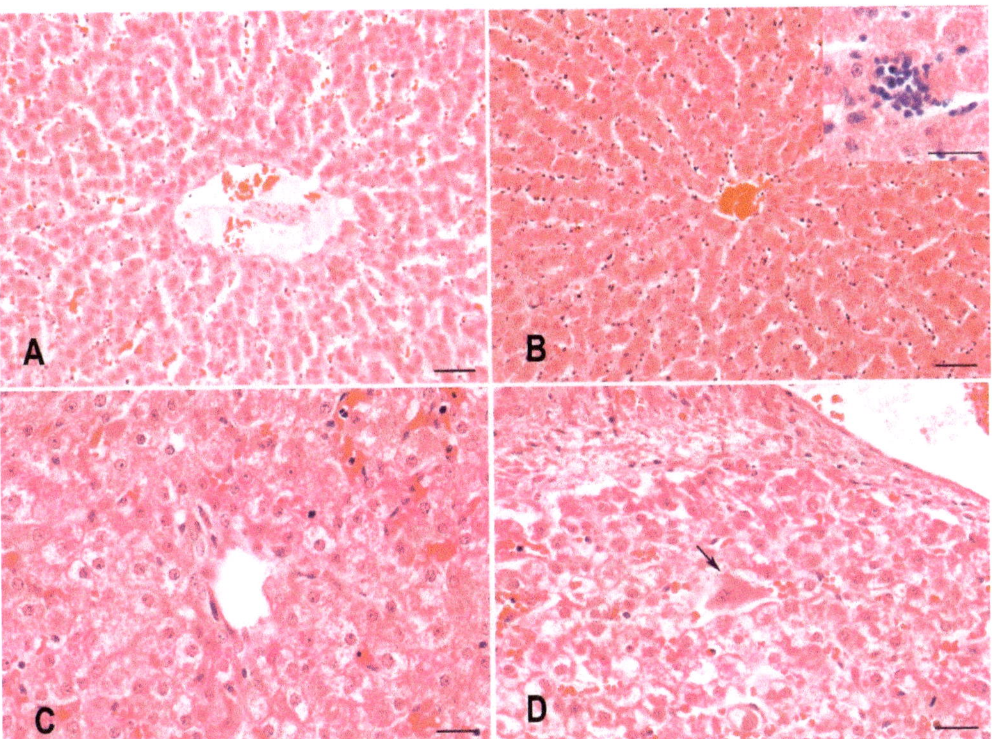

Figure 1. Representative microphotographs of the liver of gilts and their piglets: (**A**) gilt from the control group, (**B**) gilt from the experimental group with increased sinusoidal leukocytosis with inflammatory infiltrates of hepatic lobules (inset depicting the latter), (**C**) piglet from the control group with hepatocytes diffusely exhibiting vacuolar or granular cytoplasm, (**D**) piglet from the experimental group with hepatocytes diffusely exhibiting vacuolar or granular cytoplasm and a single hepatocyte exhibiting megalocytosis and megakaryosis (arrow). H&E, bar = 50 μm.

Several assessed histopathological changes, namely, irregularity of hepatic cords, biliary hyperplasia, dilatation and thickening of blood vessels, and thrombosis of blood or lymphatic vessels, did not occur in any of the liver samples from either the gilts or their piglets.

3.2. Apoptosis and Proliferation Index in the Liver of Gilts and Their Suckling Piglets

The cumulative number of apoptotic hepatocytes was significantly higher in the experimental group of gilts compared with the control group ($p = 0.0224$) and an even more significant difference was observed for the apoptotic cells in hepatic sinusoids ($p = 0.0007$) (Figure 2A,B). Moreover, the apoptotic cells in hepatic sinusoids were marginally significantly but strongly correlated with sinusoidal leukocytosis with inflammatory infiltrates of hepatic lobules (Spearman's $\rho = 0.69$, $p = 0.0535$). The apoptotic cells in hepatic sinusoids were most likely lymphocytes, based on their morphology, but double immunohistochemical labelling was not attempted to further clarify this. There was no statistically significant difference in the proliferation index of hepatocytes between the two groups of gilts ($p = 0.6901$) (Table 3).

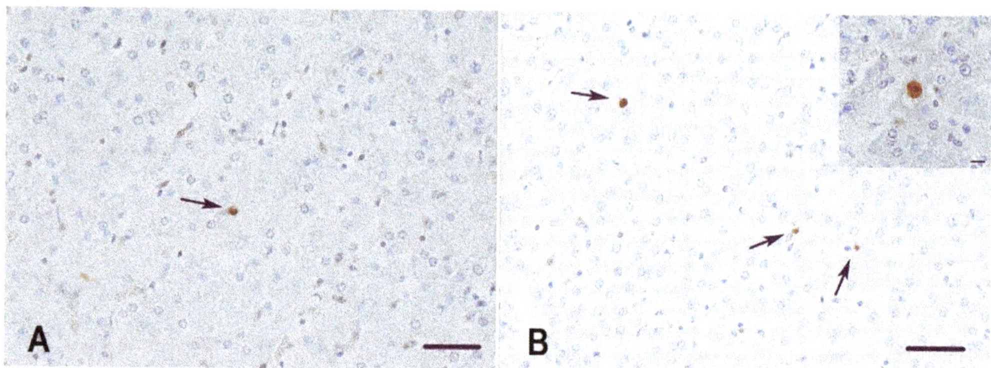

Figure 2. Representative microphotographs of apoptosis in the liver of gilts: (**A**) gilt from the control group with a single apoptotic cell in a hepatic sinusoid (arrow), (**B**) gilt from the experimental group with increased numbers of apoptotic cells in hepatic sinusoids (arrows) and an apoptotic hepatocyte (inset). TUNEL, counterstained with Mayer's hematoxylin, bar = 50 μm.

Table 3. The cumulative number of apoptotic cells and the proliferation index of hepatocytes in the liver of control and experimental gilts presented with basic descriptive statistics.

	Group	N	Min	Q_1	Median	Q_3	Max
Apoptotic hepatocytes *	control	10	0	0	0	1	2
	experimental	9	0	1	1	2	5
Apoptotic cells in hepatic sinusoids *	control	10	0	2.5	5.5	7.5	13
	experimental	9	11	13	16	19	35
Proliferation index of hepatocytes	control	10	0	0	0	0.08	0.2
	experimental	10	0	0	0	0	0.2

* Statistically significant difference between the control and experimental groups ($p < 0.05$). N—number of animals, Min—minimum, Q_1—lower quartile, Q_3—upper quartile, Max—maximum.

In piglets, there were no statistically significant differences in the cumulative number of apoptotic hepatocytes ($p = 0.1265$) and apoptotic cells in hepatic sinusoids ($p = 0.8581$) between the two groups. No statistically significant difference in the proliferation index of hepatocytes was seen when comparing both groups of piglets ($p = 0.1069$) (Table 4). Nevertheless, the proliferation index of hepatocytes was found to be strongly correlated with hepatocytes with vacuolar or granular cytoplasm (Spearman's ρ = 0.74, $p = 0.006$).

Table 4. The cumulative number of apoptotic cells and the proliferation index of hepatocytes in the liver of control and experimental piglets presented with basic descriptive statistics.

	Group	N	Min	Q_1	Median	Q_3	Max
Apoptotic hepatocytes	control	9	0	0	1	1	2
	experimental	9	0	0	0	0	1
Apoptotic cells in hepatic sinusoids	control	9	1	5	5	8	22
	experimental	9	1	4	6	6	20
Proliferation index of hepatocytes	control	9	0.2	0.3	0.3	0.6	1.2
	experimental	9	0	0.1	0.1	0.3	5.3

N—number of animals, Min—minimum, Q_1—lower quartile, Q_3—upper quartile, Max—maximum.

3.3. Interlobular Connective Tissue in the Liver of Gilts

Morphometrical analysis of interlobular connective tissue was only implemented on liver samples from both groups of gilts. Subjectively, the interlobular connective tissue in the control group of gilts appeared to be of the expected thickness, whereas in the experimental group, it appeared mildly decreased; therefore, automated detection was important to decrease subjective bias. The interlobular connective tissue in the experimental group appeared to have decreased amounts of collagen fibers, therefore forming narrower bands of fibrous connective tissue among hepatic lobules (Figure 3). The amount of interlobular connective tissue proved to be significantly lower in the experimental group compared with the control group of gilts ($p = 0.0232$) (Figure 4).

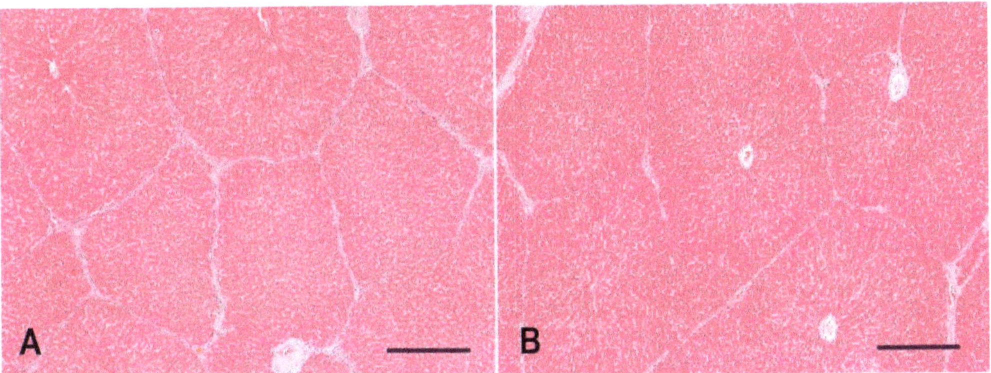

Figure 3. Representative microphotographs of hepatic lobules and surrounding interlobular connective tissue of gilts: (**A**) gilt from the control group with an expected amount of interlobular connective tissue and (**B**) gilt from the experimental group with a decreased amount of interlobular connective tissue in comparison with the control group. Goldner's Masson trichrome stain, bar = 100 µm.

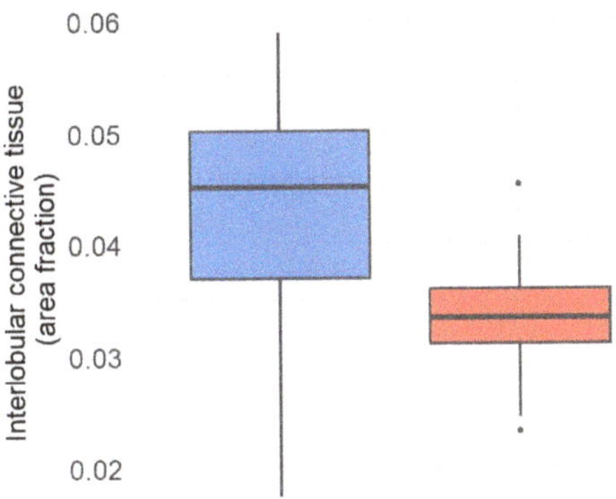

Figure 4. Distribution of interlobular connective tissue in the liver of the control and experimental groups of gilts with a statistically significant difference.

4. Discussion

Mycotoxin co-contamination of finished pig feed is more common than single mycotoxin contamination [13], emphasizing the importance of investigating the effects of such naturally contaminated feeds, even when mycotoxins are detected below the accepted tolerated concentrations for individual mycotoxins. A review from Escrivá et al. [17] showed that in vivo toxicity studies of *Fusarium* mycotoxins had become much more frequent in the decade between 2003 and 2014, thus highlighting their importance. These mycotoxicoses can manifest as acute diseases with high morbidity and death or as chronic diseases reduced animal productivity, and decreased resistance to pathogens [45].

In the present study, gilts were fed a diet containing maize naturally contaminated with *Fusarium* mycotoxins from roughly the beginning of the last quarter of their pregnancy until a week after the weaning of their piglets (a total of 54 ± 1 day). Our experimental diet contained three *Fusarium* mycotoxins, namely DON (5.08 mg per kg of feed), ZEN (0.09 mg per kg of feed), and FA (21.6 mg per kg of feed). Undertaking this approach, we assumed that the presenting liver pathology of both gilts and their piglets would mimic chronic exposure to mycotoxins in a typical pig production setting.

A detailed investigation into liver histopathology revealed a statistically significant difference in hepatocellular necrosis and apoptosis as well as sinusoidal leukocytosis with inflammatory infiltrates of hepatic lobules. Additionally, these two histopathological changes were strongly correlated, and they likely represented recruitment of inflammatory cells to sites of hepatocellular necrosis. Since the inflammatory cells were composed of a mixed population of neutrophils, eosinophils, lymphocytes, and, occasionally, macrophages, the observed rise in sinusoidal leukocytes could also be an indicator of a concurrent inflammatory process, knowing that DON is suspected to raise the susceptibility to infection and chronic diseases [46].

In humans and animals, the toxic effects of DON include emesis and anorexia, alteration of intestinal and immune functions, reduced absorption of nutrients, and elevated susceptibility to infection and chronic diseases [46]. As DON can induce such a variety of toxic effects, it is difficult to interpret whether liver pathology observed in the in vivo experiments was mostly due to direct hepatotoxic effects or significantly aggravated by indirect effects related to reduced nutrient absorption and overall decreased food intake.

Some studies have shown no changes in liver morphology when DON was either the sole potential toxic factor or in combination with another mycotoxin [30,31,40,42,43], but that was not the case in all studies. When piglets received diets either mono-contaminated with DON (1.5 mg per kg of feed) or multi-contaminated with DON (2 or 3 mg per kg of feed), ZEN (1.5 mg per kg of feed), and NIV (1.3 mg per kg of feed), the most prominent histopathological features were disorganization of hepatic cords, cytoplasmic vacuolization of hepatocytes, and megalocytosis. Piglets fed the co-contaminated diet with the higher dose of DON also exhibited focal necrosis in the liver [33]. In previous studies, feed co-contaminated with DON and ZEN often elicited microscopic changes in the liver. Chen et al. [28] fed pigs a diet containing 1 mg of DON and 0.250 mg of ZEN per kg of feed and mentioned blood vessel thickening and dilation as the only histopathological finding in the liver but did not provide a clear grading scheme or statistical analysis. When feeding prepuberal gilts for 35 days with wheat containing increasing concentrations of DON and ZEN, the amount of intracytoplasmic glycogen decreased in a dose-dependent manner, whereas hemosiderin deposition increased. Wheat containing the highest doses of DON (6.1 mg or 9.57 mg per kg of feed) and ZEN (0.235 mg or 0.358 mg per kg of feed) also elicited a statistically significant increase in the thickness of interlobular connective tissue [27]. On the other hand, a similar study, where pregnant sows were exposed to a concentration of 4.42 mg DON and 0.048 mg ZEN per kg of feed for 35 days, did not show any difference in thickness of interlobular connective tissue [29], whereas a slight difference was observed in pregnant sows receiving a concentration of 9.57 mg DON and 0.358 mg ZEN per kg of feed for 35 days [30]. Another study assessed the effects of feeding gilts for 1, 3, or 6 weeks with either DON at a dose of 0.012 mg/kg body weight (BW) per day, ZEN at 0.04 mg/kg BW, or a mixture of DON and ZEN. Histologically, several changes were observed in the liver, especially in gilts receiving DON and ZEN, such as increased thickness of perilobular connective tissue, increased total microscopic liver score, dilation of hepatic sinusoids, temporary changes in glycogen content, and increased iron accumulation in hepatocytes [34].

Interestingly, a statistically significant decrease in interlobular connective tissue was observed in our experimental group of gilts when automated detection was used on slides stained with Goldner's Masson trichrome stain. Our finding is in contrast with previous studies because in those cases, experimental animals had an increased amount of interlobular connective tissue in the liver [27,30,34]. A recent study on collagen and elastin content in skin of mink receiving DON-contaminated feed showed a decrease in type III (immature) collagen when mink received DON at a concentration of 1.1 mg/kg of feed and complete absence of type III (immature) collagen when mink received a dose of 3.7 mg/kg DON in feed with or without 0.05% bentonite [47]. As interlobular connective tissue in pig liver contains both type I and type III collagen [48], the observed decrease in interlobular connective tissue in our study may have been due to DON influencing the expression of fibrous collagens. Further investigation would be needed to confirm or refute this assumption.

Cell culture experiments on porcine hepatocytes have shown that DON causes morphological and functional disorders in hepatocytes. Cell death of hepatocytes occurred in a dose-dependent manner and exhibited morphological changes characteristic of apoptosis. Apoptosis was further confirmed by consistently TUNEL-positive nuclei and increased activity of caspase-3, a key enzyme in apoptotic cell death [49]. Our study showed a significantly increased number of apoptotic cells in the liver of experimental gilts; an increase was observed for hepatocytes and even more so for cells in hepatic sinusoids. The apoptotic cells in hepatic sinusoids were most likely lymphocytes, based on their morphology, but we did not attempt double immunolabelling to confirm this. Similarly, piglets intravenously injected with DON at a concentration of 1 mg/kg body weight displayed systemic apoptosis of lymphocytes in lymphoid tissues as well as hepatocytes, thereby proposing a hepatotoxic potential of DON next to its already known immunotoxic effect [41]. Based on these findings, we suspect that both oral and intravenous administration of DON can cause apoptosis of hepatocytes as well as circulating leukocytes.

As DON has also been associated with antiproliferative activity [50], we assessed the proliferation activity by determining the number of Ki-67-positive hepatocytes, but found no significant differences, neither in the liver of gilts nor their piglets. The same was observed in the liver of porcine fetuses when pregnant sows were exposed to DON and ZEN for 35 days during pregnancy [40], whereas an increased proliferation index was observed when piglets received feed mono-contaminated with FBs (6 mg per kg of feed) or DON (3 mg per kg of feed) and especially when co-contaminated with both FBs and DON [14].

Fusarium mycotoxins clearly have direct and indirect effects on the liver of pigs that are fed contaminated diets, thereby some studies have examined possible placentar transfer from sows to their offspring [20,21,29–31,39,40]. Some did not identify any changes in the fetuses [40] or piglets [39]. Dänicke et al. [20] suggested that the developing fetus is exposed to DON, ZEN, and their metabolites when sows are fed contaminated feed, but did not observe any histopathological changes, and Goyarts et al. [21] also found that DON and de-epoxy-DON pass the placental barrier to a significant extent. Fetuses that were exposed to DON and ZEN between the 35th and 70th day of gestation exhibited increased glycogen content and changes in the architecture of hepatocellular mitochondria, likely caused by diaplacentar toxin transfer of ingested toxins from the mother to the developing fetuses. Feed consumption did not play a role in that experiment because both the control and experimental groups received the same amount of feed per day [29]. When sows were fed a diet contaminated with DON (9.57 mg per kg of feed) and ZEN (0.358 mg per kg of feed) between the 75th and 110th day of gestation, there were no histological changes in the livers of their piglets [30]. When sows received DON and ZEN co-contaminated diet between the 63rd and 70th day of gestation, DON was detected in fetus plasma and there was a change in the proportion of their white blood cells [31]. In our study, there were no significant differences in the assessed histopathological changes, apoptosis, and proliferation index in the livers of piglets. This suggests that a decreased feed consumption by the gilts leading to a lower energy and nutrient intake or a direct effect of diaplacentar transfer of ingested mycotoxins did not cause morphologically apparent changes in the developing fetuses, possibly due to the relatively low concentrations of *Fusarium* mycotoxins in the diet fed to the pregnant gilts.

5. Conclusions

The present study showed that gilts fed a diet contaminated with DON, ZEN, and FA showed significantly increased hepatocellular necrosis and apoptosis as well as sinusoidal leukocytosis with mixed inflammatory infiltrates of hepatic lobules. The number of apoptotic hepatocytes and apoptotic cells in hepatic sinusoids was also significantly higher in the experimental gilts compared with the control gilts. No significant differences were observed in the livers of their piglets, suggesting that the herein utilized concentrations of *Fusarium* mycotoxins do not have detrimental effects on the liver of offspring.

Author Contributions: Conceptualization, B.J.-S., T.Š. and T.D.; methodology, B.J.-S., T.Š. and T.D.; formal analysis, T.K.; investigation, T.D. and T.Š.; resources, B.J.-S. and M.G.; writing—original draft preparation, T.D.; writing—review and editing, T.D., T.Š., T.K., M.G., B.L. and B.J.-S.; visualization, T.Š., T.K. and T.D.; supervision, T.Š. and B.J.-S.; project administration, B.J.-S.; funding acquisition, B.J.-S. and M.G. All authors have read and agreed to the published version of the manuscript.

Funding: This research was funded by the Slovenian Research Agency, grant number P4-0092.

Institutional Review Board Statement: The study was conducted according to the guidelines of the Declaration of Helsinki and approved by the Veterinary Administration of the Republic of Slovenia (323-02-170/2002).

Informed Consent Statement: Not applicable.

Data Availability Statement: The data presented in this study are available on request from the corresponding author.

Acknowledgments: We acknowledge the assistance of veterinary technicians Benjamin Cerk and Jurij Omahen for cutting sections from FFPE liver samples and preparing H&E and Goldner's Masson trichrome-stained slides for reviewing.

Conflicts of Interest: The authors declare no conflict of interest.

References

1. Alshannaq, A.; Yu, J.H. Occurrence, toxicity, and analysis of major mycotoxins in food. *Int. J. Environ. Res. Public Health* **2017**, *14*, 632. [CrossRef] [PubMed]
2. Wan, L.Y.; Turner, P.C.; El-Nezami, H. Individual and combined cytotoxic effects of fusarium toxins (deoxynivalenol, nivalenol, zearalenone and fumonisins B1) on swine jejunal epithelial cells. *Food Chem. Toxicol.* **2013**, *57*, 276–283. [CrossRef] [PubMed]
3. D'Mello, J.P.F.; Placinta, C.M.; Macdonald, A.M.C. Fusarium mycotoxins: A review of global implications for animal health, welfare and productivity. *Anim. Feed Sci. Technol.* **1999**, *80*, 183–205. [CrossRef]
4. Bertero, A.; Moretti, A.; Spicer, L.J.; Caloni, F. Fusarium molds and mycotoxins: Potential species-specific effects. *Toxins* **2018**, *10*, 244. [CrossRef]
5. D'Mello, J.P.F. Contaminants and toxins in animal feeds. In *FAO Animal Production and Health Paper, Assessing Quality and Safety of Animal Feeds*; Daya Publishing House: Rome, Italy, 2004; Volume 160, pp. 107–128.
6. D'Mello, J.P.F.; Porter, J.K.; Macdonald, A.M.C.; Placinta, C.M. Fusarium mycotoxins. In *Handbook of Plant and Fungal Toxicants*; D'Mello, J.P.F., Ed.; CRC Press: Boca Raton, FL, USA, 1997; pp. 287–301.
7. Gruber-Dorninger, C.; Novak, B.; Nagl, V.; Berthiller, F. Emerging mycotoxins: Beyond traditionally determined food contaminants. *J. Agric. Food Chem.* **2017**, *65*, 7052–7070. [CrossRef]
8. Lee, H.J.; Ryu, D. Advances in mycotoxin research: Public health perspectives. *J. Food Sci.* **2015**, *80*, T2970–T2983. [CrossRef]
9. Bennett, J.W.; Klich, M. Mycotoxins. *Clin. Microbiol. Rev.* **2003**, *16*, 497–516. [CrossRef]
10. Placinta, C.M.; D'Mello, J.P.F.; Macdonald, A.M.C. A review of worldwide contamination of cereal grains and animal feed with fusarium mycotoxins. *Anim. Feed Sci. Technol.* **1999**, *78*, 21–37. [CrossRef]
11. Streit, E.; Naehrer, K.; Rodrigues, I.; Schatzmayr, G. Mycotoxin occurrence in feed and feed raw materials worldwide: Long-term analysis with special focus on Europe and Asia. *J. Sci. Food Agric.* **2013**, *93*, 2892–2899. [CrossRef]

12. Streit, E.; Schatzmayr, G.; Tassis, P.; Tzika, E.; Marin, D.; Taranu, I.; Tabuc, C.; Nicolau, A.; Aprodu, I.; Puel, O.; et al. Current situation of mycotoxin contamination and co-occurrence in animal feed—Focus on Europe. *Toxins* **2012**, *4*, 788–809. [CrossRef]

13. Khoshal, A.K.; Novak, B.; Martin, P.G.P.; Jenkins, T.; Neves, M.; Schatzmayr, G.; Oswald, I.P.; Pinton, P. Co-occurrence of DON and emerging mycotoxins in worldwide finished pig feed and their combined toxicity in intestinal cells. *Toxins* **2019**, *11*, 727. [CrossRef] [PubMed]

14. Grenier, B.; Loureiro-Bracarense, A.P.; Lucioli, J.; Pacheco, G.D.; Cossalter, A.M.; Moll, W.D.; Schatzmayr, G.; Oswald, I.P. Individual and combined effects of subclinical doses of deoxynivalenol and fumonisins in piglets. *Mol. Nutr. Food Res.* **2011**, *55*, 761–771. [CrossRef]

15. Commission of the European Communities. Commission recommendation of 17 August 2006 on the presence of deoxynivalenol, zearalenone, ochratoxin A, T-2 and HT-2 and fumonisins in products intended for animal feeding. *Off. J. Eur. Union* **2006**, *L229*, 7–9.

16. European Commission. Commission Recommendation (EU) 2016/1319 of 29 July 2016 amending recommendation 2006/576/EC as regards deoxynivalenol, zearalenone and ochratoxin A in pet food. *Off. J. Eur. Union* **2016**, *L 208*, 58–60.

17. Escrivá, L.; Font, G.; Manyes, L. In Vivo toxicity studies of fusarium mycotoxins in the last decade: A review. *Food Chem. Toxicol.* **2015**, *78*, 185–206. [CrossRef]

18. Schelstraete, W.; Devreese, M.; Croubels, S. Comparative toxicokinetics of fusarium mycotoxins in pigs and humans. *Food Chem. Toxicol.* **2020**, *137*, 111140. [CrossRef] [PubMed]

19. Coffey, M.T.; Hagler, W.M.; Jones, E.E.; Cullen, J.M. Interactive effects of multiple mycotoxin contamination of swine diets. *Biodeterior. Res.* **1990**, *3*, 117–128. [CrossRef]

20. Dänicke, S.; Brüssow, K.P.; Goyarts, T.; Valenta, H.; Ueberschär, K.H.; Tiemann, U. On the transfer of the fusarium toxins deoxynivalenol (DON) and zearalenone (ZON) from the sow to the full-term piglet during the last third of gestation. *Food Chem. Toxicol.* **2007**, *45*, 1565–1574. [CrossRef]

21. Goyarts, T.; Dänicke, S.; Brüssow, K.P.; Valenta, H.; Ueberschär, K.H.; Tiemann, U. On the transfer of the fusarium toxins deoxynivalenol (DON) and zearalenone (ZON) from sows to their fetuses during days 35–70 of gestation. *Toxicol. Lett.* **2007**, *171*, 38–49. [CrossRef] [PubMed]

22. Jakovac-Strajn, B.; Vengušt, A.; Pestevšek, U. Effects of a deoxynivalenol-contaminated diet on the reproductive performance and immunoglobulin concentrations in pigs. *Vet. Rec.* **2009**, *165*, 713–718. [PubMed]

23. Malovrh, T.; Jakovac-Strajn, B. Feed contaminated with fusarium toxins alter lymphocyte proliferation and apoptosis in primiparous sows during the perinatal period. *Food Chem. Toxicol.* **2010**, *48*, 2907–2912. [CrossRef]

24. Friend, D.W.; Thompson, B.K.; Trenholm, H.L.; Boermans, H.J.; Hartin, K.E.; Panich, P.L. Toxicity of T-2 toxin and its interaction with deoxynivalenol when fed to young pigs. *Can. J. Anim. Sci.* **1992**, *72*, 703–711. [CrossRef]

25. Harvey, R.B.; Edrington, T.S.; Kubena, L.F.; Elissalde, M.H.; Casper, H.H.; Rottinghaus, G.E.; Turk, J.R. Effects of dietary fumonisin B1-containing culture material, deoxynivalenol-contaminated wheat, or their combination on growing barrows. *Am. J. Vet. Res.* **1996**, *57*, 1790–1794.

26. Tiemann, U.; Brüssow, K.P.; Jonas, L.; Pöhland, R.; Schneider, F.; Dänicke, S. Effects of diets with cereal grains contaminated by graded levels of two fusarium toxins on selected immunological and histological measurements in the spleen of gilts. *J. Anim. Sci.* **2006**, *84*, 236–245. [CrossRef] [PubMed]

27. Tiemann, U.; Brüssow, K.P.; Küchenmeister, U.; Jonas, L.; Kohlschein, P.; Pöhland, R.; Dänicke, S. Influence of diets with cereal grains contaminated by graded levels of two fusarium toxins on selected enzymatic and histological parameters of liver in gilts. *Food Chem. Toxicol.* **2006**, *44*, 1228–1235. [CrossRef]

28. Chen, F.; Ma, Y.; Xue, C.; Ma, J.; Xie, Q.; Wang, G.; Bi, Y.; Cao, Y. The combination of deoxynivalenol and zearalenone at permitted feed concentrations causes serious physiological effects in young pigs. *J. Vet. Sci.* **2008**, *9*, 39–44. [CrossRef] [PubMed]

29. Tiemann, U.; Brüssow, K.P.; Küchenmeister, U.; Jonas, L.; Pöhland, R.; Reischauer, A.; Jäger, K.; Dänicke, S. Changes in the spleen and liver of pregnant sows and full-term piglets after feeding diets naturally contaminated with deoxynivalenol and zearalenone. *Vet. J.* **2008**, *176*, 188–196. [CrossRef]

30. Tiemann, U.; Brüssow, K.P.; Dannenberger, D.; Jonas, L.; Pöhland, R.; Jäger, K.; Dänicke, S.; Hagemann, E. The effect of feeding a diet naturally contaminated with deoxynivalenol (DON) and zearalenone (ZON) on the spleen and liver of sow and fetus from day 35 to 70 of gestation. *Toxicol. Lett.* **2008**, *179*, 113–117. [CrossRef]

31. Goyarts, T.; Brüssow, K.P.; Valenta, H.; Tiemann, U.; Jäger, K.; Dänicke, S. On the effects of the fusarium toxin deoxynivalenol (DON) administered per Os or intraperitoneal infusion to sows during days 63 to 70 of gestation. *Mycotoxin Res.* **2010**, *26*, 119–131. [CrossRef]

32. Chaytor, A.C.; See, M.T.; Hansen, J.A.; de Souza, A.L.; Middleton, T.F.; Kim, S.W. Effects of chronic exposure of diets with reduced concentrations of aflatoxin and deoxynivalenol on growth and immune status of pigs. *J. Anim. Sci.* **2011**, *89*, 124–135. [CrossRef]

33. Gerez, J.R.; Pinton, P.; Callu, P.; Grosjean, F.; Oswald, I.P.; Bracarense, A.P. Deoxynivalenol alone or in combination with nivalenol and zearalenone induce systemic histological changes in pigs. *Exp. Toxicol. Pathol.* **2015**, *67*, 89–98. [CrossRef]

34. Skiepko, N.; Przybylska-Gornowicz, B.; Gajęcka, M.; Gajęcki, M.; Lewczuk, B. Effects of deoxynivalenol and zearalenone on the histology and ultrastructure of pig liver. *Toxins* **2020**, *12*, 463. [CrossRef]

35. Ujčič-Vrhovnik, I.; Švara, T.; Malovrh, T.; Jakovac-Strajn, B. The effects of feed naturally contaminated with fusarium mycotoxins on the thymus in suckling piglets. *Acta Vet. Hung.* **2020**, *68*, 186–192. [CrossRef]

36. Alm, H.; Brüssow, K.P.; Torner, H.; Vanselow, J.; Tomek, W.; Dänicke, S.; Tiemann, U. Influence of fusarium-toxin contaminated feed on initial quality and meiotic competence of gilt oocytes. *Reprod. Toxicol.* **2006**, *22*, 44–50. [CrossRef] [PubMed]
37. Bracarense, A.P.; Lucioli, J.; Grenier, B.; Drociunas Pacheco, G.; Moll, W.D.; Schatzmayr, G.; Oswald, I.P. Chronic ingestion of deoxynivalenol and fumonisin, alone or in interaction, induces morphological and immunological changes in the intestine of piglets. *Br. J. Nutr.* **2012**, *107*, 1776–1786. [CrossRef]
38. Reddy, K.E.; Jeong, J.Y.; Lee, Y.; Lee, H.J.; Kim, M.S.; Kim, D.W.; Jung, H.J.; Choe, C.; Oh, Y.K.; Lee, S.D. Deoxynivalenol- and zearalenone-contaminated feeds alter gene expression profiles in the livers of piglets. *Asian-Australas J. Anim. Sci.* **2018**, *31*, 595–606. [CrossRef] [PubMed]
39. Friend, D.W.; Thompson, B.K.; Trenholm, H.L.; Hartin, K.E.; Prelusky, D.B. Effects of feeding deoxynivalenol (DON)-contaminated wheat diets to pregnant and lactating gilts and on their progeny. *Can. J. Anim. Sci.* **1986**, *66*, 229–236. [CrossRef]
40. Wippermann, W.; Heckmann, A.; Jäger, K.; Dänicke, S.; Schoon, H.A. Exposure of pregnant sows to deoxynivalenol during 35–70 days of gestation does not affect pathomorphological and immunohistochemical properties of fetal organs. *Mycotoxin Res.* **2018**, *34*, 99–106. [CrossRef]
41. Mikami, O.; Yamaguchi, H.; Murata, H.; Nakajima, Y.; Miyazaki, S. Induction of apoptotic lesions in liver and lymphoid tissues and modulation of cytokine mRNA expression by acute exposure to deoxynivalenol in piglets. *J. Vet. Sci.* **2010**, *11*, 107–113. [CrossRef] [PubMed]
42. Stanek, C.; Reinhardt, N.; Diesing, A.K.; Nossol, C.; Kahlert, S.; Panther, P.; Kluess, J.; Rothkötter, H.J.; Kuester, D.; Brosig, B.; et al. A chronic oral exposure of pigs with deoxynivalenol partially prevents the acute effects of lipopolysaccharides on hepatic histopathology and blood clinical chemistry. *Toxicol. Lett.* **2012**, *215*, 193–200. [CrossRef]
43. Renner, L.; Kahlert, S.; Tesch, T.; Bannert, E.; Frahm, J.; Barta-Böszörményi, A.; Kluess, J.; Kersten, S.; Schönfeld, P.; Rothkötter, H.J.; et al. Chronic DON exposure and acute LPS challenge: Effects on porcine liver morphology and function. *Mycotoxin Res.* **2017**, *33*, 207–218. [CrossRef] [PubMed]
44. R Core Team. *R: A Language and Environment for Statistical Computing*; R Foundation for Statistical Computing: Vienna, Austria, 2019; Available online: https://www.R-project.org/ (accessed on 26 February 2020).
45. Antonissen, G.; Martel, A.; Pasmans, F.; Ducatelle, R.; Verbrugghe, E.; Vandenbroucke, V.; Li, S.; Haesebrouck, F.; Van Immerseel, F.; Croubels, S. The impact of fusarium mycotoxins on human and animal host susceptibility to infectious diseases. *Toxins* **2014**, *6*, 430–452. [CrossRef]
46. Payros, D.; Alassane-Kpembi, I.; Pierron, A.; Loiseau, N.; Pinton, P.; Oswald, I.P. Toxicology of deoxynivalenol and its acetylated and modified forms. *Arch. Toxicol.* **2016**, *90*, 2931–2957. [CrossRef]
47. Taszkun, I.; Tomaszewska, E.; Dobrowolski, P.; Żmuda, A.; Sitkowski, W.; Muszyński, S. Evaluation of collagen and elastin content in skin of multiparous minks receiving feed contaminated with deoxynivalenol (DON, vomitoxin) with or without bentonite supplementation. *Animals* **2019**, *9*, 1081. [CrossRef] [PubMed]
48. Nishimura, S.; Sagara, A.; Oshima, I.; Ono, Y.; Iwamoto, H.; Okano, K.; Miyachi, H.; Tabata, S. Immunohistochemical and scanning electron microscopic comparison of the collagen network constructions between pig, goat and chicken livers. *Anim. Sci. J.* **2009**, *80*, 451–459. [CrossRef] [PubMed]
49. Mikami, O.; Yamamoto, S.; Yamanaka, N.; Nakajima, Y. Porcine hepatocyte apoptosis and reduction of albumin secretion induced by deoxynivalenol. *Toxicology* **2004**, *204*, 241–249. [CrossRef] [PubMed]
50. Nagashima, H. Deoxynivalenol and nivalenol toxicities in cultured cells: A review of comparative studies. *Food Saf.* **2018**, *6*, 51–57. [CrossRef] [PubMed]

MDPI

St. Alban-Anlage 66

4052 Basel

Switzerland

Tel. +41 61 683 77 34

Fax +41 61 302 89 18

www.mdpi.com

Animals Editorial Office

E-mail: animals@mdpi.com

www.mdpi.com/journal/animals